THE MAN WHO WASN'T THERE

THE MAN WHO WASN'T THERE

INVESTIGATIONS INTO THE STRANGE NEW SCIENCE OF THE SELF

Anil Ananthaswamy

DUTTON
est. 1852

DUTTON
— est. 1852 —

An imprint of Penguin Random House LLC
375 Hudson Street
New York, New York 10014

Portions of chapters 1, 7, and 8 appeared in *New Scientist* magazine. Most of chapter 3 was first published as a feature in the online magazine *Matter.* The parable narrated in the prologue is adapted with permission from Jonardon Ganeri. The English translation of the parable appears in his book *The Self.*

LIBRARY OF CONGRESS CATALOGING-IN-PUBLICATION DATA
Ananthaswamy, Anil.
The man who wasn't there : investigations into the strange new science of the self / Anil Ananthaswamy.
pages cm
Includes index.
ISBN 978-0-525-95419-4 (hardback)
1. Neuropsychology. 2. Identity (Psychology) 3. Mind and body. I. Title.
QP360.A48 2015
616.80092'2—dc23 2015003576

Printed in the United States of America
10 9 8 7 6 5 4 3 2 1

Set in Warnock Pro
Designed by Daniel Lagin

to

those of us who want to let go
but wonder, who *is letting go of* what?

CONTENTS

It seems outlandish that the centerless universe, in all its spatio-temporal immensity, should have produced me, of all people . . . There was no such thing as me for ages, but with the formation of a particular physical organism at a particular place and time, suddenly there *is* me, for as long as the organism survives. . . . How can the existence of one member of one species have this remarkable consequence?

—Thomas Nagel

THE MAN WHO WASN'T THERE

PROLOGUE

An allegory about a man who was devoured by ogres first appears in an ancient Indian Buddhist text of the *Madhyamika* (the middle-way) tradition. It dates from sometime between 150 and 250 CE and is a somewhat gruesome illustration of the Buddhist notion of the true nature of the self.

A man on a long journey to a distant land finds a deserted house and decides to rest for the night. At midnight, an ogre turns up carrying a corpse. He sets the corpse down next to the man. Soon, another ogre in pursuit of the first arrives at the deserted house. The two ogres begin bickering over the corpse. Each claims to have brought the dead man to the house and wants ownership of it. Unable to resolve their dispute, they turn to the man who saw them come in, and ask him to adjudicate. They want an answer. Who brought the corpse to the house?

The man, realizing the futility of lying to the ogres—for if one won't kill him, the other one will—tells the truth: the first ogre came with the corpse, he says. The angry second ogre retaliates by ripping off the man's arm. What ensues gives the allegory its macabre twist. The first ogre immediately detaches an arm from the corpse and attaches it to the man. And so it goes: the second ogre rips a body part off the man; the first ogre replaces it by taking the same body part from the corpse and attaching it to the man. They end up swapping everything—arms, legs, the torso, and even the head. Finally, the two ogres make a meal of the corpse, wipe their mouths clean, and leave.

The man, whom the ogres have left behind, is extremely disturbed. He is left pondering what he has witnessed. The body that he was born in has been eaten by the ogres. His body now is made up of body parts of someone else entirely. Does he now have a body or doesn't he? If the answer is yes, is it his body or someone else's? If the answer is no, then what is he to make of the body that he can see?

The next morning, the man sets off on the road, in a state of utter confusion. He finally meets a group of Buddhist monks. He has a burning question for them: does he exist or does he not? The monks throw the question back at him: who are you? The man is not sure how to answer the question. He's not sure he's even a person, he says—and tells the monks of his harrowing encounter with the ogres.

What would modern neuroscientists tell the man if he were to ask them *Who am I?* While some would likely point out the near-biological implausibility of what the ogres did, they would nonetheless have some tantalizing answers. These answers, which strive to illuminate the "I," are the focus of this book.

1

THE LIVING DEAD

WHO IS THE ONE WHO SAYS, "I DON'T EXIST"?

Men ought to know that from the brain, and from the brain only, arise our pleasures, joys, laughters and jests, as well as our sorrows, pains, griefs and tears. . . . These things that we suffer all come from the brain. . . . Madness comes from its moistness.

—Hippocrates

If I try to seize this self of which I feel sure, if I try to define and to summarize it, it is nothing but water slipping through my fingers.

—Albert Camus

Adam Zeman will never forget the phone call. It was, as he called it, a "*Monty Python*–esque" summons from a psychiatrist, asking him to come urgently to the psychiatric ward. There was a patient who was claiming to be brain dead. Zeman felt as if he were being called to the intensive care unit, not the psychiatric ward. Yet, "this was very unlike the kind of call you normally receive from the ICU," Zeman told me.

The patient, Graham, was a forty-eight-year-old man. Following a separation from his second wife, Graham had become deeply depressed and had tried to kill himself. He got into his bath and pulled an electric heater into the bathwater, wanting to electrocute himself. Fortunately, the fuse blew and Graham was spared. "It didn't seem to have done any physical damage to him, but some weeks later he formed the belief that his brain had died," said Zeman, a neurologist at the University of Exeter in the UK.

It was a rather specific belief. And one that led Zeman to have some very strange conversations. "Look, Graham, you are able to hear me, see me, and understand what I'm saying, remember your past, and express yourself, surely your brain must be working," Zeman would say to Graham.

Graham would say, "No, no, my brain's dead. My mind is alive but my brain is dead."

Worse yet, Graham was distraught at his unsuccessful attempt at suicide. "He was one of the undead or half-dead," Zeman told me. "He in fact went and spent quite a bit of his time, for a while, in graveyards, because he felt he was with his own when he was there."

Zeman quizzed Graham to understand the grounds for this belief. It became clear that something very fundamental had shifted. Graham's subjective experience of himself and his world had changed. He no longer felt he needed to eat or drink. Things that once used to give him pleasure no longer did. "When he pulled on a cigarette, nothing happened," Zeman told me. Graham claimed that he never needed to sleep, that he did not feel sleepy. Of course, he was doing all of these things—eating, drinking, sleeping—but his desire for these and the intensity of his feelings had damped down dramatically.

Graham had lost something we all have: a keen sense of our own appetites and emotions. Patients suffering from depersonalization often report this emotional dulling or flatness. Depression too can bring about similar states of being, where emotions lose their edge. But these patients don't go on to develop such stark delusions of nonexistence. In Graham's case, the loss of emotional vividness was so extreme that "he had come to the conclusion, on the basis of that alteration in experience, that his brain must have died," said Zeman.

Zeman thinks that two key factors play a role in such robust delusions. One is a profound alteration in the quality of one's sense of oneself and the world—in Graham's case, the emotional rug had been pulled from under his feet. The second is an alteration in one's ability to reason about that experience. "Both things seemed to be true in Graham's case," said Zeman.

Graham's delusion was immune to evidence to the contrary. Zeman, in his conversations, would bring Graham to the point of surrender—to make him see the falseness of his delusion. Graham would acknowledge that a whole range of his mental faculties was intact, that he could see, hear, speak, think, remember, and so on.

So Zeman would say to him, "Clearly, Graham, your mind is alive."

He'd say, "Yeah, yeah, the mind is alive."

"The mind has a lot to do with the brain; surely your brain is alive," Zeman would prod him.

But Graham would not take the bait. "He'd say, 'No, my mind is alive, but my brain is dead. It died in that bath,'" Zeman told me. "You could get quite close to producing what you would think was knockdown evidence, but he wouldn't accept it." It was intriguing that Graham had developed such an explicit delusion—that of being dead because his brain was dead. Would his delusion have been different in an era when the legal definition of death did not include brain death?

Over the course of his medical practice, Zeman had only ever seen one other case of someone claiming to be dead. In the mid-1980s, working as a junior doctor in Bath, England, Zeman had to treat a woman who had undergone protracted bowel surgery and was suffering from severe malnourishment. Her body had been ravaged by repeated surgery. "She became very depressed as a result of that and formed a belief that she had died," Zeman said. "Which in a strange way seemed understandable to me, because the kind of trauma she was undergoing was so awful. She thought she was dead."

Zeman recognized the symptoms in Graham, and diagnosed him as suffering from Cotard's syndrome, which was first identified as a distinct disorder by the nineteenth-century French neurologist and psychiatrist Jules Cotard.

● ● ○

Walk down the rue de l'École-de-Médecine in the Sixth Arrondissement in Paris, and you'll see a formidable colonnade. A striking example of French neoclassical architecture, the colonnade forms a portico for the Université René Descartes. Designed in the late eighteenth century by architect Jacques Gondouin, the façade, as the architect intended, demands attention and yet feels open and inviting.

I entered the building to visit the rare manuscript section of the Library of the School of Medicine, to look at a document on the life of Jules Cotard. The document is the text of a eulogy delivered by his friend and colleague Antoine Ritti in 1894, almost five years after Cotard's death. Cotard had been devotedly nursing his daughter, who was suffering from diphtheria, but then fell ill himself with the disease and died in 1889. Much of what we know of Cotard comes from Ritti's eulogy, a copy of which exists amid the pages of an old leather-bound volume, whose spine simply reads *MÉLANGES BIOGRAPHIQUES*—a mixture of biographies. I turned the pages to Ritti's eulogy. Handwritten on the first page was a note to the then head of the faculty of medicine of the university: "*Hommage de profond respect,*" the note read. It was signed *Ant. Ritti.*

Cotard is best known for describing what are called nihilistic delusions, or *délire des négations*. But before he came up with that phrase, Cotard first talked of "delirium in a severely melancholic hypochondriac" at a meeting of the Société Médico-Psychologique on June 28, 1880, using as an example the case of a forty-three-year-old woman who claimed "she had 'no brain, nerves, chest, or entrails, and was just skin and bone,' that 'neither God or the devil existed,' and that she did not need food, for 'she was eternal and would live forever.' She had asked to be burned alive and had made various suicidal attempts."

Soon afterward, Cotard coined the phrase *délire des négations*, and after his death, other doctors named the syndrome after him. Over time, "Cotard's delusion" has come to refer to the most striking symptom of the syndrome—the belief that one is dead. However, the syndrome itself refers to a constellation of symptoms, and does not have to include the delusion of being dead or not existing. The other symptoms include the belief that various body parts or organs are missing or putrefying, feelings of guilt, feelings of being damned or condemned, and paradoxically, even feelings of immortality.

But it's the delusion that one does not exist that poses an interesting philosophical challenge. Until recently, the seventeenth-century French philosopher René Descartes's assertion *Cogito ergo sum* (I think, therefore I am) was the bedrock of Western philosophy. Descartes established a clear dualism of mind and body: the body was of the physical world, something that takes up space and exists in time, while the mind's essence was thought and it did not extend into space. For Descartes, *cogito* did not mean thinking as much as "clear and distinct intellectual perception, independent of the senses." An implication of Descartes's philosophy, according to philosopher Thomas Metzinger, was that "one cannot be wrong about the contents of one's own mind."

This Cartesian idea has been falsified in many disorders, including Alzheimer's, where patients are often unaware of their own condition. Cotard's syndrome is also a puzzle. Metzinger argues that we should be paying attention to what it feels like to be suffering from Cotard's—what philosophers call the *phenomenology* of a disorder. "Patients may explicitly state not only that they are dead, but also that they don't exist at all." While this seems logically impossible—an obviously alive individual claiming not to exist—it *is* part of the phenomenology of Cotard's.

I left the library, and stepped back out onto the rue de l'École-de-Médecine, and turned around to take another look at the name "Université René Descartes" etched into the stone above the colonnade. There was something intriguing about researching Jules Cotard in a university named after Descartes. What does Cotard's eponymous delusion say about Cartesian ideas? Is the Cotard's syndrome patient saying, "I think, therefore *I'm not*"?

● ● ○

"Who is the I that knows the bodily me, who has an image of myself and a sense of identity over time, who knows that I have propriate strivings? I know all these things, and what is more, I know that I know them. But who is it who has this perspectival grasp?"

Who, indeed. The American psychologist Gordon Allport's lyrical musings above capture the central conundrum of being human. We instinctively and intimately know what he's referring to. It is there when we wake up and slips away when we fall asleep, maybe to reappear in our dreams. It is that feeling we have of being anchored in a body we own and control, and from within which we perceive the world. It is the feeling of personal identity that stretches across time, from our first memories to some imagined future. It is all of these tied into a coherent whole. It is our sense of self. Yet, despite this personal intimacy we have with ourselves, elucidating the nature of the self remains our greatest challenge.

All through recorded history, it is clear that humans have been fascinated and confounded by the self. Pausanias, a Greek traveler during Roman rule, wrote about the maxims inscribed at the fore-temple at Delphi by seven wise sages. One maxim said, "Know thyself." The Kena Upanishad, among the most analytical and metaphysical of Hindu scriptures, begins with these words: "By whom commanded and directed does the mind go towards its objects? . . . At whose will do men utter speech? What power directs the eye and the ear?"

Saint Augustine said this of the notion of time, but he might as well have been speaking about the self: "If no one asks of me, I know; if I wish to explain to one who asks, I know not."

And so it is that from the Buddha to the modern neuroscientist and philosopher, humans have pondered the nature of the self. Is it real or an illusion? Is the self in the brain, and if so, where in the brain is it? Neuroscience is telling us that our sense of self is an outcome of complex interactions between brain and body, of neural processes that update the self moment by moment, the moments strung together to give us a seamless feeling of personhood. We often hear of how the self is an illusion, that it is nature's most sophisticated sleight of hand. But all this talk of tricks and illusions obfuscates a basic truth: remove the self and there is no "I" on whom a trick is being played, no one who is the subject of an illusion.

● ● ○

From the Université Rene Descartes, it's a thirty-minute walk down rue des Écoles, past the national museum of natural history, to reach the Pitié-Salpêtrière Hospital, where Jules Cotard started his medical career as an intern in 1864. I went there to see David Cohen, the head of the hospital's infant and adolescent psychiatry unit.

Over the course of his medical residency and practice, Cohen has seen a few handfuls of patients who have suffered from Cotard's syndrome. Given the rarity of this disorder, this relatively large sample has given Cohen an intimate look at Cotard's. We talked of one particular patient, fifteen-year-old May—one of the youngest recorded cases of Cotard's. Cohen treated her and had extensive discussions with her after she recovered, enabling him to link her delusions with her personal history. He got a peek into how the self, even in a delusional state like Cotard's, is influenced by one's personal narrative and even dominant cultural norms.

About a month before May came to Cohen's clinic, she had started feeling extremely sad and depressed, and eventually began exhibiting delusions about her own existence. By the time she was admitted, she had become severely catatonic—mute and unmoving. "Even the nurses were terrified by her," Cohen told me. But with a few days of inpatient psychiatric care, May recovered somewhat, just enough to say a few words each day, which the nurses would write down diligently. Between these sporadic intimations from May and discussions with her parents, Cohen pieced together May's story.

Her family was middle-class Catholic. May had two siblings, a brother and a sister. The sister, who was ten years older, had married a dentist. The family had a history of depression: their mother had suffered from severe depression before May was born, and one of May's aunts had undergone electroconvulsive therapy (ECT), which involves delivering mild pulses of electricity to the brain to induce seizures, and is often an effective treatment for severe depression—though almost always of last resort.

May's delusions were classic Cotard's. "She was telling us that she had no teeth, no uterus, and that she had this feeling of being already dead," Cohen said. He struggled to describe May's condition in English. "I don't know the word in English . . . *morts vivants!*" he said. I looked it up later: the literal translation is *the living dead*.

"She was waiting to be buried . . . in a coffin," said Cohen.

When her condition didn't improve even after six weeks of therapy and medication, Cohen suggested ECT. Given the family's experience with depression, her parents immediately agreed. After six treatments, May appeared to recover, so Cohen stopped the ECT—but she relapsed immediately, prompting Cohen to resume the treatment. This time she did recover, except for some headache, mild confusion, and slightly disturbed memory. When she began talking, it was as if she had awakened from a nightmare.

Cohen's discussions with her—in which he asked May to talk freely of any associations that came to mind when he mentioned her delusions—shed surprising light. For instance, the delusion that she had no teeth seemed to have something to do with her sister's husband, the dentist. Cohen discerned that she may have had feelings for her brother-in-law. She spoke of never wanting to be treated by him. Again, Cohen struggled for the correct word in English to describe the way she expressed herself. *Pudique*, he said in French. "Modest." She spoke of her brother-in-law in "such a way that you understood that she'll never be naked in front of him."

Her delusions about missing her uterus seemed to be tied to episodes of masturbation. "She felt very guilty about that and she thought that maybe she would be sterile."

Cohen was making the point that the specificity of the delusions is related to one's autobiography and the cultural context. To make his case for the latter, he recalled a fifty-five-year-old man who had come to see him in the 1990s. Cohen diagnosed him with Cotard's. One of his delusions was that he had AIDS—which he didn't. Cohen figured his delusion was linked to guilt over his hypersexuality during the manic phase of his bipolar disorder, from which he also suffered. Before the 1970s, hypochondriac delusions in Cotard's patients, if they involved sexually transmitted diseases, were almost always related to syphilis—the cultural scourge of the times. Interestingly, this man had actually contracted syphilis while serving in the military as a young man (Cohen tested him for antibodies to confirm). But his delusions during his Cotard's episode, which happened decades later, were not about syphilis but HIV/AIDS—which had supplanted syphilis in the broader culture as "God's punishment for sins of the flesh" (syphilis almost never shows up anymore during hypochondriac delusions in Cotard's). "It's only one case, [but] I think this case is very informative," Cohen said.

For Cohen, Cotard's syndrome is revealing of the workings of the self. The disorder is a deeply felt disturbance of one's being, and shows that the self is linked to one's body, one's story, and one's social and cultural milieu. Brain, body, mind, self, and society are inextricably linked.

● ● ○

Back in Exeter, Adam Zeman had encountered something similar with Graham. The delusion in Graham's case was that his mind was alive but his brain was dead. "It was an updated, contemporary version of the Cotard's delusion. To come to the conclusion that your brain has died in isolation, . . . [you need] a concept of brain death, which is a relatively recent medical development."

What Zeman found even more intriguing was the inherent dualism in Graham's delusion—that an "immaterial" mind can exist independent of the brain and the body. "I thought it rather beautifully illustrated the dualism to which most of us are prone," Zeman told me. "The idea that your mind could be alive while your brain is dead is a rather extreme expression of dualism."

Philosophical musings aside, Zeman found Graham's situation sad. "He was slow and flat, with very little emotional modulation in his voice. [I] occasionally got a flicker of a smile, but there was rather little facial expression," said Zeman. "You had the sense of someone for whom existence was extremely bleak, and for whom thought was something of an effort."

● ● ○

A patient suffering from Cotard's syndrome is often extremely depressed. A depression far more serious than most of us can understand. I was given an insight into this by yet another French psychiatrist, William de Carvalho, whom I also met in Paris—at his office on avenue Victor-Hugo. He drew me a line diagram to illustrate where Cotard's stands on the depression scale. He started with "normal" on the left, then added "sad," "depressed," "very depressed," "melancholic" at equal intervals on the right. Then he added a series of dots—the progression was not linear anymore—and at the end of those dots he wrote, "Cotard's." "With Cotard's there is like a great black wall that goes from Earth to Saturn. You can't look over it," said de Carvalho, a dapper man of French-Senegalese descent with a way with words.

He had a private practice but also worked at the renowned Sainte Anne Hospital in Paris. He remembered one Cotard's patient that he treated in the early 1990s who showed classic signs of "melancholic omega." The phrase has its origins in Charles Darwin's descriptions of melancholia in his book *The Expression of Emotions in Man and Animals*: "a facial expression involving a wrinkling of the skin above the nose and between the eyebrows that resembles the Greek letter omega." While Darwin wrote about these "grief muscles" on the face, it was German psychiatrist Heinrich Schüle who coined the term "melancholic omega" in 1878, based on Darwin's vivid descriptions.

Dr. de Carvalho's patient was a fifty-year-old engineer and poet. The man had faked trying to kill his wife—he put his hands around her neck, then stopped, and told her to call the police. When the police came, they saw a very disturbed, even bizarre, man. So they took him directly to Sainte Anne Hospital rather than the police station (the man's act had a copycat quality to it: in 1980, the French philosopher Louis Althusser, who had been suffering from depression, strangled his wife, and was taken to a mental hospital first instead of being sent to jail).

The day after the incident, de Carvalho met the man at Sainte Anne Hospital. "I asked him, 'Why are you trying to kill your wife?' He said, 'Well, it's such a crime that I deserve to [have] my head cut.' He was hoping that he would be killed, even [though] there was no death penalty in France."

The man was exhibiting an extreme form of another symptom characteristic of Cotard's syndrome: guilt. "He told me at the time that he was worse than Hitler. And he asked us to help him to be killed, because he was so bad for humanity," said de Carvalho.

The patient had lost weight, his beard was unkempt and overgrown, and he had stopped bathing because he felt he had no right to take showers and use up too much water. The hospital decided to make a film about him (for their archives) while he was still in the throes of Cotard's. At one point in the filming, the patient pulled a white sheet over his head. "I'm so bad, I don't want people looking at that film to be touched by such badness," he told de Carvalho, who was behind the camera. Dr. de Carvalho pointed out that it was just a film, he couldn't possibly affect anyone through it. "And he said, 'I know, but it's like that; I am so bad,'" de Carvalho told me. Also, the broader culture had again influenced the man's delusion. He was convinced he was responsible for the AIDS epidemic and that people would get AIDS just by watching the film.

Many months later, after the man had recovered (the treatment included ECT), de Carvalho watched the film with his former patient. At the end of the twelve-minute film, the man turned to de Carvalho and said, "Well, this is very interesting. But who is it?" De Carvalho thought the man was joking.

"That's you," de Carvalho told the man.

"No, it's not me," the man replied.

Soon, de Carvalho realized that there was no point trying to convince him. He was not the same man as the one who had descended into the darkness that is Cotard's.

Given such extreme depression during Cotard's, psychiatrists have wondered why most sufferers don't attempt suicide. Partly, it's because the patients are unable to act, like deer caught in headlights. But de Carvalho thinks they don't attempt suicide because they feel they are already dead. "And you can't be more dead than dead."

● ● ○

When Zeman began talking with Graham and realized the extent of his depression and delusion, he suspected an underlying neurological cause. Something had altered Graham's sense of self and perception of his environment. There was one neurologist who would know what to look for: Steven Laureys at the University of Liège in Belgium. Zeman took Graham's consent and sent him to Liège with a community psychiatric nurse in tow. Graham reached the university hospital in Liège and asked for Dr. Laureys.

The secretary called. Laureys, like Zeman, will never forget the phone call: "Doctor, I have a patient here who is telling me he is dead. Please come over."

● ● ○

Many of the patients Laureys sees are in a bad way. Some are comatose, some in a state of unresponsive wakefulness (previously called vegetative), others are minimally conscious, and yet others are people suffering from locked-in syndrome (those who are conscious but completely paralyzed, and are sometimes able to move only their eyes).

After more than a decade of work with such patients as well as healthy subjects, Laureys's team has identified a network of key brain regions in the frontal lobe (the part of the cortex beneath the forehead) and the parietal lobe (which is behind the frontal lobe). He considers activity in this network to be the signature of conscious awareness. This awareness can be analyzed in two dimensions, he told me. One is awareness of the external world: everything you perceive through your senses, whether it's vision, touch, smell, sound, or taste. The other dimension is internal awareness, something more closely related to the self, which includes the internal perception of one's body, thoughts that are triggered regardless of external stimuli, mental imagery and daydreaming, much of which is self-referential. "It's an oversimplification to reduce this very rich complexity we call consciousness, but I think it's meaningful to take those two dimensions," he emphasized.

And indeed, Laureys's team has shown that the frontoparietal network associated with conscious awareness is actually two different networks. Activity in one correlates with awareness of the external: a network of lateral frontoparietal brain areas—the regions on the outer side of the frontal and parietal lobes. The other correlates with awareness of the internal and is potentially related to aspects of the self: a network of areas along the brain's midline—the inner parts of the frontal and parietal lobes, near the cleft that separates the two hemispheres of the brain.

Studies in healthy patients showed that these two dimensions of awareness are inversely correlated: if you are paying attention to the external world, then activity in the network associated with external awareness goes up while the regions associated with internal awareness dampen down. And vice versa.

Besides this frontoparietal network, there's another key region of the brain that's involved in conscious awareness: the thalamus.

There are long-distance two-way connections between the thalamus and the frontoparietal network, and Laureys's work suggests that it's the dynamics of information exchange and processing in these regions that takes us from being merely aroused to being consciously aware.

However, throughout our discussion, Laureys repeatedly insisted, "We should not be neo-phrenologists." He was referring to the dubious field of phrenology pioneered by German doctor Franz Joseph Gall (1758–1828), who argued that each and every mental faculty was the product of a specific brain region, and that these regions created characteristic bumps on the skull. So you could, in theory, run your fingers over someone's skull and figure out the relative strength of these "organs" inside their brain.

The self, said Laureys, is not something that can be localized to one brain area.

● ● ○

When Laureys met Graham, he too found Graham a very depressed man. Laureys noticed Graham's blackened teeth; he had stopped brushing. Graham repeated the same story that he told Adam Zeman—that he was brain dead. "He was not faking anything," Laureys told me. "So we scanned him."

Did he object to being scanned? I asked.

"He said 'I don't care,'" said Laureys.

Despite his condition, Graham was still using the first-person pronoun, "I," to refer to himself.

Laureys's team produced both magnetic resonance imaging (MRI) and positron emission tomography (PET) scans of Graham's brain. The MRI scans showed no structural brain damage. But the PET images revealed something very interesting: the frontoparietal network associated with external and internal conscious awareness had very low metabolic activity. Part of the internal awareness network is the so-called default mode network (DMN), which has been shown to be active during self-referential activity. A key hub in this network is a brain region called the precuneus—one of the most connected regions in the brain. In Graham's case, the default mode network and the precuneus were far too quiet—almost down to levels Laureys had seen in patients in a state of unresponsive wakefulness. It's true that Graham was on medication, but Laureys thinks that medication alone could not explain the extent of the lowered metabolism.

The lowered metabolism had also spread to the lateral surface of the frontal lobes—specifically some regions that are known to be involved in rational thought.

Though both Laureys and Zeman cautioned against making too much out of one case, the results are suggestive. It's likely that the impaired metabolic activity in the midline regions had caused Graham to have an altered self-experience—maybe a greatly reduced sense of self. But because that lowered metabolic activity had spread to other regions of the frontal lobes, he was unable to talk himself out of that altered experience, as he otherwise might have. He became convinced he was brain dead.

A more recent case study, published in November 2014, also supports this hypothesis. Two Indian doctors were treating a sixty-five-year-old woman with dementia, when she began to show signs of classic Cotard's. "Our patient presented to us with beliefs like 'I think I am dead and what I am is not me,' 'I do not exist,' 'there is nothing in my brain, just vacuum,' and 'it is infectious and I'm infecting my close relatives and I am responsible for all their suffering,'" Sayantanava

Mitra of the Sarojini Naidu Medical College, Agra, India, wrote to me in an email.

Mitra's team scanned her and the MRI scan revealed that the frontotemporal brain regions had atrophied. They noted, in particular, that a deep-brain region called the insula was heavily damaged in both hemispheres. There's growing evidence that the insula is responsible for the subjective perception of our body states, a crucial aspect of our conscious experience of selfhood. So, a damaged insula was likely hampering the woman's sense of her own body, and her dementia made it difficult for her to correct false perceptions, leading to claims of being dead.

The doctors started her on mild doses of antipsychotic and antidepressant medications. She recovered enough to take part in psychotherapy, with the therapist using her MRI scans as "evidence against her belief that her head was rotten," Mitra said. The therapist was able to shake her out of her false beliefs. She was eventually discharged, and continues to get better on her medication.

Graham, too, eventually recovered. Cotard's syndrome is, thankfully, transient in most people, even though the treatment at times might involve electroconvulsive therapy.

"I think Cotard's delusion is a victory of metaphor over simile," Zeman told me. "There are mornings when most of us get up and feel as if we are half-dead. So alterations of your experience which you might express using that kind of simile are not so uncommon. But the bizarre thing about Cotard's is that people begin to treat this simile as if it were literally true. And for that to happen, there surely has to be some disturbance of reason."

● ● ○

The paucity of patients with Cotard's syndrome means that the neural underpinnings of their delusions are yet to be fully understood, but it's clear that Cotard's syndrome is giving us a glimpse into the nature of the self.

Take, for instance, something philosopher Shaun Gallagher calls the immunity principle, an idea that goes back to Austrian philosopher Ludwig Wittgenstein. It refers to the fact that when we make a statement like "I think the Earth is flat," we can be wrong about Earth's flatness, but we cannot be wrong about the "I," the subjective self that is making the assertion. When we use the pronoun "I," the word refers to the one who is the subject of an experience, not someone else. I cannot be wrong about that. Or can I?

Cotard's delusion certainly gets philosophers thinking (if they need any further enticement), as do various other conditions, such as schizophrenia. In Cotard's delusion, the firm belief that "I don't exist" seemingly challenges the immunity principle. But even though the delusional person is wrong about the nature of his existence (which is analogous to Earth's flatness), the immunity principle holds because there is still an "I" making the claim, and that "I" cannot refer to anyone else but the person experiencing nonexistence.

What or who is that "I"? The question permeates this book. Whoever or whatever the "I," it manifests itself as a subject of experiences.

But how does the brain, with its physical, material processes, give rise to a seemingly immaterial, private mental life (at the core of which seems to be the "I," the subjectivity)? This is the so-called hard problem of consciousness. Neuroscience doesn't have an answer so far. Philosophers disagree vehemently on whether science can ever solve this problem, or whether this problem is illusory, one that might disappear as we understand the brain in more and more detail. This book does not offer neuroscientific solutions to the hard problem of consciousness—there are none, yet.

But this book does address the nature of the self. One way to think of the self is to consider its many facets. We are not just one thing to others or even to ourselves; we present many faces. The great American psychologist William James identified at least three such facets: the material self, which includes everything I consider as me or mine; the social self, which depends on my interactions with others ("a man has as many social selves as there are individuals who recognize him and carry an image of him in their mind"); and the spiritual self ("a man's inner or subjective being, his psychic faculties or dispositions").

The search for the self is also well served by thinking of it in terms of two categories: the "self-as-object" and the "self-as-subject." It turns out that some aspects of the self are objects to itself. For instance, if you were to say, "I am happy"—the feeling of happiness, which is part of your sense of self at that moment, belongs to the self-as-object category. You are aware of it as a state of your being. But the "I" that feels happy—the one that is aware of its own happiness—that's the more slippery, elusive self-as-subject. The same "I" could also be depressed, ecstatic, and anything in between.

With this distinction in mind, if you take Laureys's studies, which show that in healthy subjects the frontoparietal network activity constantly switches back and forth from internal to external awareness, what seems to be changing is the content of one's consciousness: from awareness of external stimuli to awareness of aspects of one's self. When you are self-aware, in that you are conscious of your own body, your memories, and your life story, aspects of the self become the contents of consciousness. These comprise the self-as-object.

It's possible that parts of this self-as-object are not being experienced vividly in Cotard's syndrome. Whatever it is that tags objects in our consciousness as mine or not-mine, self or not-self, may be malfunctioning (we'll see in coming chapters some mechanisms that could be behind such tagging). In Graham's case, the *mineness* or vividness that is usually attributed to, say, one's body and/or emotions was maybe lacking. And the resulting untenable belief that he was brain dead entered his conscious awareness unchallenged, given his underactive, low-functioning lateral frontal lobes.

But regardless of what one is aware of, isn't there someone who is always the subject of the experience? Even if you are completely absorbed in something external, say, a melancholic violin solo—and the contents of your consciousness are devoid any self-related information, whether of your body or worries about your job—does the feeling that *you* are having that experience ever go away?

To help us get closer to some answers, we can turn to insights of people suffering from various perturbations of the self, which serve as windows to the self. Each such neuropsychological disorder illuminates some sliver of the self, one that has been disturbed by the disorder, resulting at times in a devastating illness.

These words from Lara Jefferson's *These Are My Sisters: A Journal from the Inside of Insanity* leave us in no doubt of the damage wrought to the self in a schizophrenic person: "Something has happened to me—I do not know what. All that was my former self has crumbled and fallen together and a creature has emerged of whom I know nothing. She is a stranger to me. . . . She is not real—she is not I . . . she is I—and because I still have myself on my hands, even if I am a maniac, I must deal with me somehow."

But in the devastation are clues to what makes us who we are. These maladies are to the study of the self what brain lesions are to study of the brain: They are cracks in the façade of the self that let us examine an otherwise almost impenetrable, ongoing, unceasing neural process. And while what follows in the coming chapters is not an exhaustive list of all neuropsychological conditions that disturb the self, I have chosen conditions that satisfied at least two criteria: first, they were amenable to studying some distinct aspect of the self, and second, there is significant ongoing science that specifically addresses these conditions from the perspective of the self.

In Alzheimer's we get a sense of one's story unraveling. If you can't answer the question "Who am I?" with declarative statements ("I am Richard," "I am a retired professor," and so on), either because your memory is failing you or the brain regions that let you reflect upon these characteristics are damaged, have you lost your sense of self? If so, have you lost all of it, or part of it? What if, despite the cognitive disintegration of your coherent story—what some call the narrative or autobiographical self—other aspects of you still functioned?

Ralph Waldo Emerson is thought to have suffered from Alzheimer's. He also wrote eloquently about memory and its role in making us who we are. But Emerson was curiously indifferent about his own dementia. It's one of the characteristics of Alzheimer's disease that sufferers are sometimes unaware of their own condition. Alzheimer's was the unmaking of his identity, including identifying himself as diseased.

The next chapter examines Alzheimer's and its role in the undoing of a person, while asking: is some essence of selfhood—despite a ravaged brain in the late stages of disease—preserved in the body? The celebrated American composer Aaron Copland (1900–1990) also suffered from Alzheimer's disease. At times he wouldn't know where he was, but he could still conduct his signature orchestral suite *Appalachian Spring.* Who or what swung the conductor's baton?

Body integrity identity disorder—a curious condition in which people feel that some part of their body, usually limbs, is not their own, often leading them to the horrific act of severing the body part—gives us a glimpse into how the brain constructs a sense of one's own body, the bodily self.

Schizophrenia can fragment a person—and part of this fragmentation is due to a compromised sense of agency, the feeling we all have that we are the agents of our actions. What if this feeling—a crucial aspect of the self—goes awry? Could it lead to psychosis?

Then there is depersonalization disorder, which robs the self of its emotional substrate, making us strangers to ourselves, thus highlighting the role of emotions and feelings in creating the self.

Autism sheds light on the developing self. Children with autism are usually unable to instinctively "read" others' minds, which then leads to problems relating socially to others, but is this ability also tied to reading one's own mind and hence self-awareness? There's tantalizing new work suggesting that the roots of this impairment lie in an autistic brain's inability to make sense of the body and its interactions with the environment, leading first to an uncertain bodily self and then to behavioral problems.

Out-of-body experiences and the more complex doppelgänger effect (in which people perceive and interact with a duplicate of their own body) reveal that even the most basic things we take for granted—being grounded in a body, identifying with it, and viewing the world from behind our eyes—can be disrupted, thus giving us a glimpse of the components necessary for a low-level self that potentially precedes all else.

Ecstatic epilepsy begets a condition that borders on the mystical, when we are truly here and now, fully aware of our own being, yet paradoxically bereft of boundaries, leading to a feeling of transcendental oneness. Is this condition bringing us closer to the essence of the self—a self that maybe endures for just moments and is at the heart of the debate about whether there is or there isn't a self?

We conclude with a journey to Sarnath, India, where the Buddha, nearly 2,500 years ago, gave his first sermon. Buddhist ideas of no-self seem to resonate with what some modern philosophers are saying about the self—that it's illusory. But is it really? Does empirical evidence support the idea that the self is a made-up entity? Insights gleaned from the maladies of the self will help us make sense of age-old questions and maybe even ask a few of our own.

● ● ○

While visiting David Cohen in Paris, I asked him about May, his fifteen-year-old Cotard's syndrome patient. "Who is it that is saying she doesn't exist?"

"This is the mystery of psychiatry," Cohen said. "We always say that there is something . . . that can still relate to the real world, even in the most crazy state."

In Liège, Steven Laureys's PhD student Athena Demertzi, who helped Laureys scan and study Graham, told me something about Graham that also reminded me that despite his delusion of being brain dead, there was an essence that remained. Graham had just come out of the scanner when Demertzi asked him, "Are you OK?"

"I'm OK," replied Graham.

"Alive and kicking?" she asked.

"Kicking," said Graham, pointedly.

The self is both remarkably robust and frighteningly fragile. This book, I hope, brings to life this essential paradox of who we are.

2

THE UNMAKING OF YOUR STORY

MEMORIES, A PERSON, A NARRATIVE—AND ITS UNRAVELING

Memory, connecting inconceivable mystery to inconceivable mystery, performs the impossible by the strength of her divine arms; holds together past and present,—beholding both,—existing in both . . . and gives continuity and dignity to human life. It holds us to our family, to our friends. Hereby a home is possible.

—Ralph Waldo Emerson

All those moments will be lost in time, like tears in rain.

—Replicant Roy Batty in *Blade Runner*

Allan, Michaele, and I are sitting in the living room of their home in California. Allan is settled into a large, high-backed, brown leather sofa, looking distinguished with his white beard and mustache and balding pate, and surprisingly dark eyebrows. At first glance I'm unable to tell anything's amiss. Michaele sits on a chair next to him. I ask Allan if he has any brothers or sisters. He says no, and then corrects himself immediately. "Oh, I had a brother who was demented," he says.

"Retarded," Michaele gently corrects him.

"Retarded," Allan agrees. "No one knew he was retarded until he was [about] four. I was eighteen. I didn't understand a lot."

"But you were ten when he was four," Michaele says.

"OK," says Allan.

"Allan, do you remember much about your brother?" I ask.

"A sadness about him," says Allan. "Because he couldn't talk and stuff like that. I'd take him for a walk or something like that. He never said a word."

After a small pause, he adds, "I don't even know if he's still alive."

"No, honey, he died," says Michaele. "He died the year you and I met."

Allan and Michaele met nearly thirty years ago. Allan had been a philosophy professor at a community college, Michaele a forty-year-old working as a midwife, back at school after finding herself at a cusp in her life.

"Do you remember how he died?" asks Michaele.

"I thought he died in his sleep or something," says Allan.

Actually, Allan's brother had been hospitalized for a blood clot, and while at the hospital he fell out of an upper-floor window and died. At the time, thirty years ago, Allan had told Michaele that his brother, given his diminished mental capacities, would not have had the wherewithal to jump; he had probably wanted to get home and likely stepped out of the window thinking he was on the ground floor.

When Michaele reminds Allan of this during our conversation, he says, "Oh, that's something I wanted to forget, but no . . . fell out of the window . . ." He mumbles; his words meander.

"What did they say at the hospital?" asks Michaele.

"I was too sad and too young to take it in," says Allan.

Michaele turns to me and points out that Allan was fifty years old when his brother died.

● ● ○

On December 21, 1995, researchers in Germany found a blue cardboard file that had been missing for nearly ninety years. The file contained the case report for a patient named Auguste D, a fifty-one-year-old woman from Frankfurt. A handwritten note in the file, dated November 26, 1901, captured an exchange between Auguste and her doctor, Aloysius "Alois" Alzheimer, which the German researchers published in the journal *Lancet* in 1997 (with Auguste's answers italicized):

> She sits on the bed with a helpless expression. What is your name? *Auguste.* Last name? *Auguste.* What is your husband's name? *Auguste, I think.* Your husband? *Ah, my husband.* She looks as if she didn't understand the question. Are you married? *To Auguste.* Mrs D? *Yes, yes, Auguste D.* How long have you been here? She seems to be trying to remember. *Three weeks.* What is this? I show her a pencil. *A pen.* A purse and key, diary, cigar are identified correctly. At lunch she eats cauliflower and pork. Asked what she is eating she answers *spinach.* When she was chewing meat and asked what she was doing, she answered *potatoes* and then *horseradish.* When objects are shown to her, she does not remember after a short time which objects have been shown. In between she always speaks about twins.

Three days later, Alzheimer made further notes:

> On what street do you live? *I can tell you, I must wait a bit.* What did I ask you? *Well, this is Frankfurt am Main.* On what street do you live? *Waldemarstreet, not, no....* When did you marry? *I don't know at present. The woman lives on the same floor.* Which woman? *The woman where we are living.* The patient calls *Mrs G, Mrs G, here a step deeper, she lives....* I show her a key, a pencil and a book and she names them correctly. What did I show you? *I don't know I don't know.* It's difficult isn't it? *So anxious, so anxious.* I show her 3 fingers; how many fingers? *3.* Are you still anxious *Yes.* How many fingers did I show you? *Well this is Frankfurt am Main.*

Auguste died on April 8, 1906. By then, Alzheimer had moved from Frankfurt to the Royal Psychiatric Clinic in Munich, so he had Auguste's brain sent there, where he "sampled thin slices of this brain tissue, [and] stained them with silver salts." After affixing these slices between glass slides, "Alzheimer put down his habitual cigar, removed his pince-nez, and peered into his state-of-the-art Zeiss microscope. Then, at a magnification of several hundred times, he finally saw her disease."

Summer passed and in the fall, on November 4, Alzheimer presented his findings at the 37th Conference of South-West German Psychiatrists in Tübingen. Auguste, he said, had "progressive cognitive impairment, focal symptoms, hallucinations, delusions, and psychosocial incompetence." More to the point, the cells in her cerebral cortex showed weird abnormalities.

The following year, Alzheimer published a paper called "A Characteristic Serious Disease of the Cerebral Cortex," in which he detailed the abnormalities. One was found inside neurons: "In the centre of an otherwise almost normal cell there stands out one or several fibrils due to their characteristic thickness and peculiar impregnability." Alzheimer also identified "miliary foci," places between cells where he saw aggregates of a strange substance.

It was a new form of dementia. In 1910, Emil Kraepelin, the director of the Royal Psychiatric Clinic, coined the term "Alzheimer's disease" for such strange cases of dementia, and wrote, "The clinical interpretation of this Alzheimer's disease is still unclear. Although the anatomical findings suggest that we are dealing with a particularly serious form of senile dementia, the fact is that this disease sometimes starts as early as in the late forties."

The abnormalities Alzheimer had identified in Auguste D's brain were what are now called neurofibrillary tangles and plaques of beta-amyloid protein. While neuroscientists are still debating which comes first—the neurofibrillary tangles or the beta-amyloid plaques (with some wondering whether there are precursors to these neuropathologies)—it's clear that these aberrant proteins are involved in the ruthless progression of the disease.

If Auguste D had come to see a neurologist today, she would have been diagnosed with Alzheimer's disease.

● ● ○

Michaele had been working as a lay midwife, helping with home births, in the early 1980s when midwifery was not strictly regulated in California. But with the legal issues surrounding her work becoming more challenging, Michaele decided to go back to nursing school. As part of her studies, she took a class in philosophy being taught by a charismatic fifty-year-old professor. He sauntered into the classroom wearing a leather jacket and large horn-rimmed tortoiseshell glasses, with white hair and a beard, and discussed philosophy and governments with dramatic flair. "I believe governments should be run by Romanian gypsies and ballet dancers, instead of dictators and greedy politicians," Michaele recalled him saying. She was mesmerized.

Soon, they began seeing each other ("lots of notes on the door and clandestine meetings after class," said Michaele). He was in the process of separating from his wife, and drinking a lot; she too was struggling with a bad marriage, which had started disintegrating around the time she went back to school. They both had children. But none of that stopped them as they tumbled into love.

The day I met Allan, I asked him about what Michaele had said, about being blown away by him. "Well, we were both blown the same way," he said, his voice surprisingly firm and confident. "It was . . ." He then struggled to find the right word. "Things that twirl up in the air." Tornado, I offered as a suggestion. "Tornado," he agreed.

Eventually, they bought a house together (the one I visited), got married, traveled together, often to Europe, and remade their lives around each other. Michaele recalled that one of her sons had pointed this out in his toast at their wedding: "It's always been my mom and Allan against the world . . . They have made it work . . . They have made a life for themselves, despite the challenges."

Nothing about Allan's personality had prepared Michaele for what was to come. "When it started happening, I never imagined, *never*, that he'd ever be a person with dementia," she said.

The first hints came in spring 2003. Michaele and Allan took a weekend break and went up the Eel River in Northern California, and stayed at the Benbow Historic Inn in Garberville. When they came back home on Monday, they found their answering machine overflowing with messages from Allan's department secretary and students.

Allan had completely forgotten that he had scheduled a final exam that day. It was the first serious indication that something was wrong with his memory.

In September of that year, they went to Europe for a vacation, and Michaele found Allan unable to cope with anything new. He constantly got lost, could not navigate through the French countryside, would put his ATM card into a movie-rental machine, and even had trouble packing his own suitcase.

Back in California, Allan began showing further signs of dementia. He'd forget how to get to his daughter's home, which was not too far from where they lived. There were other things that seemed amiss. "I'd come home and find him cleaning the hot tub with the circuit breaker still on, which was very dangerous. You can get an electric shock," Michaele told me. "And when I'd tell him to go turn off the circuit breaker, he would go looking in the garage, when it was on the other side of the house."

It took a year for them to see a neurologist. Allan failed some standard tests (counting backward from 100 in decrements of 7, for example, which requires the patient to concentrate and is a test for declining cognitive ability), but he still did reasonably well, which the neurologist attributed to Allan's high intelligence. The MRI suggested some small occlusions of blood vessels. The neurologist diagnosed Allan as being in the early stages of vascular dementia (the decline of cognitive processes due to impaired blood flow in the brain). A few years later, Allan's diagnosis was changed to Alzheimer's disease.

Meanwhile, Allan's personality was changing too. During their entire love affair and married life prior to his diagnosis, Allan had been a kind, sweet man. He and Michaele would have the usual arguments that all couples do, but they would resolve them almost immediately by talking things through. "He was very present," Michaele told me.

Not so once Alzheimer's set in. The slightest argument and Allan would storm out of the house, slamming the door behind him, and go "tearing off in his car." He also wrote notes incessantly—an old habit that now provided glimpses into Allan's morphing personality. "Some of them were really mean," said Michaele. "If I had called him on something, he'd go write a nasty note, 'Bitch-fest number 5.'" Or simply, "Bitch, bitch, bitch . . ."

The notes also revealed his torment at having been diagnosed with dementia. Michaele recalled one note that said "Get me out of this fucking hole." Allan read all the books he could find on the disease, and even read *Final Exit*, a book that offered those suffering from terminal illnesses a way out via assisted suicide (he kept the book on the end table near his bed). He told Michaele, "I don't want to ever end up in diapers. I don't want to ever end up in a nursing home. You have to take me down to the Bay and push me off the pier." Michaele knew she would do no such thing. "I couldn't do that, Allan, I would go to jail," she recalled telling him. "I could be charged with murder. If you want to do it, I totally get it, but you are going to have to do it. I'd support you, but I won't help you, at least not do it for you."

In the face of Alzheimer's, Allan's sharp intelligence and his cerebral personality became a double-edged sword. "The thought of him losing his brain was the most horrific thing because that's the thing he valued the most: his brain, his intelligence," Michaele said.

● ● ○

"Alzheimer's disease robs you of *who you are*. I don't think there's any greater fear for a person than to think I've lived my whole life accumulating all these memories, all these value systems, all of this place and my family and a society and here's a disease that's just going to come in and every single day just rip out the connections, just tear out the seams that actually define *who I am as a person* [italics mine]," says Rudolph Tanzi, professor of neurology at Harvard University, eloquently articulating the scary eventual outcome of Alzheimer's disease in the PBS documentary *The Forgetting: A Portrait of Alzheimer's.*

Talk at any length to caregivers like Michaele about their loved ones, and you cannot escape the conclusion that the disease destroys the very essence of one's being. At least, that's the perception from the outside.

"It's very hard," said Clare, a sixty-year-old woman of Norwegian descent living in California. "Somebody that you grew up with disappears before your eyes." Clare's ninety-year-old father is in the late stages of Alzheimer's and the family has moved him into an assisted-living center. Clare visits him often, as does her mother. "He physically looks the same but when you look in his eyes, there's nothing there," Clare told me, her voice dropping to a whisper. "There's really nothing there."

The vast medical literature will concur with Clare. Consider the phrases used to describe Alzheimer's impact: "a steady erosion of selfhood," "unbecoming" a self, "drifting towards the threshold of unbeing," and even "the complete loss of self."

Still, there are scientists, particularly social scientists, who are challenging such notions. If Alzheimer's erodes the self, does the erosion go all the way until, truly, nothing's left? We know that Alzheimer's disease destroys cognitive abilities, to the point where the person is unable to take care of himself or herself, where putting on one's pants or brushing one's teeth becomes impossible, to say nothing of the ability to recall the date and time, or recognize family members. But given that the person's sensory and motor functions are spared, does anything remain of one's self when cognition and its attendant abilities are wiped out?

Answering such questions requires us to return to what philosophers, scientists, and social scientists think the self is. Some argue that the self is fundamentally a narrative construct. It's true that one of the key aspects of the self is the narrative—the story or stories we tell others and indeed ourselves about who we are; these stories depend on remembering and imagining. "Individuals construct private and personal stories linking diverse events of their lives into unified and understandable wholes. These are stories about the self. They are the basis of personal identity and self-understanding and they provide answers to the question 'Who am I?'" wrote psychologist Donald Polkinghorne.

It's not hard to accept that our various narratives are part of the self, but is the self constituted solely of narratives, or does it have other aspects that exist before narratives are formed? Some philosophers argue that narratives constitute the self in its entirety—nothing's left once the narrative goes. To them, "the self is ultimately nothing but a dense constellation of interwoven narratives, an emergent entity that gradually unfurls from (and is thus constituted by) the stories we tell and have told about us."

This notion of narrative-as-selfhood even places the cognitive act of constructing narratives at the heart of being a self. But the experiences of people suffering from Alzheimer's pose at least two challenges to this view of the self.

One is the idea that cognition—and its role in creating narratives—is central to the self. Pia Kontos, of the University of Toronto, has been observing people with dementia for more than a decade and takes issue with such a notion of the self. "There is something of who we are that exists separate and independent from cognition," she told me.

She realizes that this claim is controversial. "It's a perspective that challenges the whole Western construction of selfhood, because central to [our understanding of selfhood] is rationality and independence and control. It sort of goes back to [Descartes], the split between mind and body. It's not just a split between mind and body, but it's a very particular notion of dualism, where the body is relegated to nothingness. It's just an empty shell, and really everything, in terms of sense of self and agency and intentionality, is attributed to the mind." Kontos wants to bring the body into the discourse on selfhood, agency, and even memory.

So, even if we were to view selfhood as a narrative, the narrative would not be solely the purview of cognition; the body has its say.

And Alzheimer's disease defies the narrative-only view of the self in yet another way. It challenges those who argue that the self is best understood as constituted of *and by* narratives—and that there is nothing else besides these narratives. While the disease does destroy one's ability to have and tell a coherent story, what remains once this narrative self disintegrates is less clear. "It is by no means obvious that . . . any experience that remains is merely an anonymous and unowned experiential episode, so that the 'subject' no longer feels pain or discomfort as his or her own," writes philosopher Dan Zahavi. Understanding what might be left behind once the narrative self is gone could point us to the brain processes that beget the self.

Zahavi, for example, argues that the self, before it becomes a full-fledged narrative, must be something minimal, something that is capable of being the subject of an experience in any given moment.

So, despite its relentless degradations, Alzheimer's disease is allowing us to examine the self in ways that are more layered and more nuanced. This emerging picture of the self will tell us what it means to be a person with end-stage Alzheimer's and influence how we take care of those people.

● ● ○

Around the time Clare's father was diagnosed with Alzheimer's disease, he walked over to the local police station—in a small town an hour's drive north of Sacramento, California—and handed the police a gun that he owned. "I got rid of the pistol because I'm afraid I'm going to use it on myself," he told Clare. Soon after, Clare's parents sold their farm and moved to a much smaller house. The farm had become too much to manage.

Clare's parents came to America from Europe when Clare was four years old. Her father was a scientist who worked for a major corporation, was extremely successful, retired early, bought his family a farm—something Clare's mother had always wanted—and settled into an altogether different kind of life amid farmers, ranchers, and cattlemen's associations. It was during one of her visits to the farm that Clare noticed something was amiss. They were having a barbecue. Her father, who always did the barbecuing, turned around to Clare and said, "I don't know quite how to do this." Clare thought he was kidding; surely he knew how to barbecue? "Oh, come on, you know what you are doing," she said. "No, Clare, I'm not [kidding]. Something is wrong," he said. It felt wrong to him in the same way that it felt wrong when he couldn't think of certain words.

He had told Clare earlier that, at times, he had to search for the right words for something he had in mind (the way Allan had struggled for the word "tornado"). Clare realized that this was a big deal for her father. "Perhaps everybody says this about their father, and especially their father with Alzheimer's disease, but in my case, I really think it was true," Clare told me. "He was an extraordinarily intelligent man. He spoke seven languages and he was quite good at them. And so words were kind of easy for him and when they started disappearing, he was concerned."

Such concerns grew. Clare remembered one key moment with her father that made it clear he was getting worse. And it centered on something he used to love: sailing. He was an accomplished sailor, who could navigate the seas at night by the stars and often would charter a large sailboat, invite friends, and sail around the Great Lakes, even the Caribbean. Once, in the 1980s, they were sailing off St. Barths (with an "interesting" group of sailors that included two Argentinians, recalled Clare; she was the only woman on board), when a massive storm hit the region. Clare's response was "Let's get out of here," but her dad (the captain) wasn't bailing out just yet. The dinghy attached to their sailboat was being whipped around by the wind and the waves and taking on water, becoming a drag on their sailboat. Everyone wanted to cut the dinghy free, but Clare's father refused to do so. Through it all he remained keenly aware of his boat; even while asleep in the middle of the night, having dropped anchor, many a time he'd know that the anchor hadn't quite grabbed. He'd make Clare get up and together they would reset the anchor. He saw them through the storm.

More than a decade later, Clare and her cousin decided to take her dad on a sailing trip. By this point, they knew something wasn't quite right with him. So this time, Clare's cousin acted as captain and Clare the first mate. Her dad was just a fellow sailor. "He knew his way around the boat, and he could sail, but he also was quite happy that he wasn't in charge. He wasn't offended or outraged by that," recalled Clare. But there were indications of his difficulties. Sometimes, when Clare and her cousin were doing a tough maneuver, her dad would stand up and risk getting hit by the swinging boom. She would have to shout out, asking him to sit down. And yet when they set a course and put him behind the wheel, he could still maintain a compass bearing, trimming the sails as necessary.

Then suddenly, out of the blue, he'd say, "What day is it? What day is it? What day is it?"

A few years later (and after he had been diagnosed with Alzheimer's), Clare and her dad were walking through the center of a small coastal town and saw a church sale. There was a model sailboat on display. Clare's father picked it up. "He looked at it, and he looked at it. He knew that he was interested in it, but I also had the sense he wasn't really sure why he was interested in it," Clare told me. "None of it was verbalized. It was in the way he was holding his body. The look on his face, which was sort of looking, but yet there was no comprehension that I could see going on." The once-accomplished sailor no longer remembered a sailboat.

She knew then that her dad's Alzheimer's had moved way beyond just short-term memory lapses.

● ● ○

Much of what we know about memory and the brain structures that support it comes from the study of one unusual man, who, for better or for worse, lived *in the moment* from the age of twenty-seven. Students of psychology and neuroscience know him as patient H. M.; he was Henry G. Molaison, born in 1926. Henry began having epileptic seizures when he was ten years old, possibly because of a minor head injury a few years earlier (but the causation was unclear; there might have been a genetic predisposition, given that Henry's cousins on his father's side also suffered from epilepsy). The seizures got progressively worse. Anticonvulsant drugs had no effect, so much so that Henry, a high-school graduate, barely managed to continue at his job on a typewriter assembly line. Eventually, in 1953, when Henry was twenty-seven years old, the neurosurgeon William Beecher Scoville of Hartford Hospital in Connecticut decided on a risky experimental surgery to treat Henry's epilepsy.

Scoville drilled two holes just above Henry's eye sockets, into which he inserted flat brain spatulas—a neurosurgeon's version of the tongue depressor—parting the frontal and temporal lobes in the two brain hemispheres. This gave him access to brain structures of the medial temporal lobe, such as the amygdala and hippocampus. He then sucked out a chunk of normal brain tissue, including much of the amygdala and hippocampus. The effect this surgery had on Henry, his name now anonymized to H. M. in the academic literature, is neuroscience lore.

H. M. continued to take anticonvulsant medication, and his grand mal seizures dropped dramatically in intensity and frequency (from once a week to once a year). But something far more intriguing happened to his memory. He "could no longer recognize the hospital staff nor find his way to the bathroom, and he seemed to recall nothing of the day-to-day events of his hospital life." In a paper published in 1957, Scoville and a psychologist at the Montreal Neurological Institute, Brenda Milner, wrote about H. M.'s psychological examination: "This was performed on April 26, 1955. The memory defect was immediately apparent. The patient gave the date as March, 1953, and his age as 27. Just before coming into the examining room he had been talking to Dr. Karl Pribram, yet he had no recollection of this at all and denied that anyone had spoken to him. In conversation, he reverted constantly to boyhood events and seemed scarcely to realize that he had had an operation."

H. M. continued to live a life lacking in new memories (a condition called anterograde amnesia), and there was also a limit to what he could recollect about his past. Milner continued to study H. M., a baton she would pass on to her student Suzanne Corkin. In 1984, Corkin wrote:

> A striking feature of H.M. is the stability of his symptoms during the 31 postoperative years. He still exhibits a profound anterograde amnesia, and does not know where he lives, who cares for him, or what he ate at his last meal. His guesses as to the current year may be off by as much as 43 years, and, when he does not stop to calculate it, he estimates his age to be 10 to 26 years less than it is. In 1982, he did not recognize a picture of himself that had been taken on his 40th birthday in 1966. Nevertheless, he has islands of remembering, such as knowing that an astronaut is someone who travels in outer space, that a public figure named Kennedy was assassinated, and that rock music is "that new kind of music we have."

H. M.'s condition highlighted the different kinds of memory we possess, some of which were intact in him, while others had been obliterated. For starters, his short-term working memory was fine; he could retain a handful of numbers for tens of seconds. But surgery had scarred some forms of H. M.'s long-term memory.

His semantic memory—the ability to remember facts and concepts—was largely intact, but only for things that he had experienced before his surgery. Meanwhile, his episodic memory, which is the memory of an episode of experience and is linked to place and time, was ruined even for his pre-surgery days. Semantic and episodic memories are forms of long-term memory called declarative or explicit memory, which requires us to consciously access information. H. M.'s anterograde amnesia was so complete that he had no declarative memory for anything that happened after his surgery (though he did manage to remember the floor plan of the house he moved into after his surgery, and lived in from 1958 to 1974; gradual accumulation of knowledge over the years, aided no doubt by the fact that he physically inhabited and moved about in the same space for years, had somehow helped H. M. form a memory of where he lived—tantalizing evidence for the body's place in forming the self in concert with the brain).

The other broad category of long-term memory is called implicit, nondeclarative, or procedural memory. This is memory that does not require conscious access. Think about knowing how to ride a bicycle. It's memory that we access unconsciously. It was Milner's classic study of H. M., published in 1962, that showed us that distinct brain structures are involved in these various types of memory. In this study, H. M. was shown two star-shaped patterns, one inside the other. He was asked to replicate the pattern by drawing between the lines of the outer and inner patterns. To complicate things further, H. M. had to draw while looking at the reflection of his hand, the pencil, and the patterns in a mirror. Amazingly, H. M. got better and better at the task over three days—while retaining absolutely no memory of having done the task. It was clear that the surgery had not messed up his procedural memory. The question was: what exactly were the brain structures that had been removed by the surgery? The papers written by Scoville after he performed H. M.'s surgery in 1953 were extremely illustrative for their time, but weren't the definitive word. In the 1990s and 2000s, H. M. underwent several brain scans, but they were, like all scans, noninvasive and hence somewhat limited in what they could precisely reveal about the excised brain regions. But more was revealed upon his death.

H. M. died on December 2, 2008. His body was transported to Mass General Hospital in Charlestown, Massachusetts, where neuroscientists spent nine hours imaging his brain. Later, a neuropathologist skillfully removed H. M.'s brain from his skull. All this led to a high-resolution 3-D model of H. M.'s brain, based on numerous fine-grained MRI scans. It was possible, finally, to dissect H. M.'s brain inside the computer. The new images confirmed what the earlier MRI scans had revealed: the back half of H. M.'s hippocampus in both hemispheres—which Scoville thought he had fully removed—was intact. But Scoville had removed something else in its entirety: the entorhinal cortex—the interface between the hippocampus and the neocortex (the part of the cortex that's unique to mammals). Alzheimer's disease begins in the entorhinal cortex and spreads. According to the literature, "It is the most heavily damaged of all cortical areas in Alzheimer's disease."

As for H. M., the "unforgettable amnesiac," he might have left no survivors, but he left behind an indelible mark on science. His profound amnesia sparked a debate over the question of whether he had a sense of self after his surgery. A similar question haunts those who are confronted with Alzheimer's disease today.

● ● ○

When most of us think of the sense of self, we are thinking of the stories in our heads about who we are. If you had to tell a story about yourself to someone else (or even to yourself) you may have to delve into your album of episodic memories that defines you. Call it the narrative self (aspects of this selfhood would not just be cognitive, but embodied, as Pia Kontos emphasizes). A narrative, by definition, is a sequence of episodes strung together. In some sense, that's what we are—a seemingly seamless narrative. As humans, we also have the ability to project this story into the future. Our narrative self, then, is not merely a remembered past but also an imagined future. Over the past decade, numerous studies have shown that the same brain networks that are responsible for remembering past events are also recruited when constructing future scenarios. For example, if you are a good sailor, as Clare's father was, you will use the same brain networks to remember last year's sailing trip as for imagining navigating the seas a few years hence. Key brain regions that form these networks include structures in the medial temporal lobe (the parts closer to the midline), including the hippocampus and the entorhinal cortex. It's these regions that are often first affected by Alzheimer's disease; it's here that the disease gains a foothold for its destructive march, eventually erasing a person's ability to construct a coherent narrative self.

In some Alzheimer's patients, this disruption of the narrative self manifests initially as anosognosia—not recognizing that you in fact have Alzheimer's. Joseph Babinski coined the term "anosognosia" in 1914 (in Greek, *agnosia* means lack of knowledge and *nosos* means disease) to describe an extremely odd behavior in some of his patients whose entire left sides were paralyzed. In his influential paper he wrote, "I want to draw attention to a mental disorder that I had the opportunity to observe . . . which consists in the fact that patients seem unaware of or ignore the existence of their paralysis." Babinski's patients not only denied or were unaware of their paralysis, they also came up with rationalizations for their lack of knowledge. Babinski wrote about one patient, "If she was asked to move her right arm, she immediately executed the command. If she was asked to move the left one, she stayed still, silent, and behaved as if the question had been put to somebody else." A particularly severe form of anosognosia is seen in Anton's syndrome (named after neuroscientist Gabriel Anton, 1858–1933), in which patients who have become blind because of damage to both sides of their occipital lobes insist that they can see.

Anosognosia in Alzheimer's can range from mild unawareness to outright denial. Neuroscientist William Jagust, an expert on Alzheimer's disease at the Lawrence Berkeley National Laboratory, has encountered the entire gamut of reactions during his years of clinical practice. "The spouse brings the patient to the doctor, and the patient says, 'Nothing's wrong with me, you are crazy,' and they have fights and all that. . . . But more often, it's that the patient doesn't really notice [the disease], is unaware, rather than [in denial]," Jagust told me. Often, after intense meetings in which Jagust would tell the patient and the family about the diagnosis, the patient would soon forget the diagnosis. It's in the nature of the disease. "After you tell them they have Alzheimer's disease, they have to stop driving, they will want to drive. The family will say, 'The doctor said you have Alzheimer's disease' and they say, 'He didn't say that!' "

Allan, too, didn't want to give up driving. Before his formal diagnosis, he began having panic attacks while driving on freeways. So he stuck to local city driving, which still worried Michaele. She'd find unexplained dents on his car, even signs of being sideswiped by another vehicle (Allan claimed the other driver was to blame, but Michaele suspected it was Allan's fault). Allan once tried to cover up an accident by spray-painting a scraped fender. A social worker at Allan's clinic warned Michaele that Allan could be sued if he got into an accident (by now he had been formally diagnosed). Allan's doctor alerted the Department of Motor Vehicles, and the DMV sent Allan a letter, asking him to come in and redo his written and road tests. "Damn it if he didn't pass the test," said Michaele. "I couldn't believe it." Eventually, to her relief, Allan's car was stolen and trashed by thieves. For Allan, it was a body blow. "He was very sad. He wrote a whole lament to his Honda on yellow paper, on just how much it meant to him, how the loss of his autonomy was a tragedy, that he wasn't a whole person anymore," Michaele told me. "It's like his sense of self was eroding."

Allan's anosognosia was perhaps mild in comparison to the level of denial of those with paralysis. Nevertheless, Alzheimer's is allowing us to understand the neural mechanisms behind anosognosia and its relationship to the sense of self. It's these mechanisms that Giovanna Zamboni, a neurologist at Oxford University, is studying. In one of her studies, she found that Alzheimer's patients with anosognosia were far better at judging traits of a close friend, caregiver, or relative than they were at judging themselves. The tasks, which were done inside an fMRI scanner, revealed that the medial prefrontal cortex (MPFC) and the left anterior temporal lobe in Alzheimer's patients were less active during self-appraisal than during tasks that required the appraisal of others (normal controls and those with mild cognitive impairment showed no such difference).

The tests reveal that anosognosia in Alzheimer's is not just a problem of memory—it's also a problem of self. "It reflects a very selective inability of updating the information regarding you, but not regarding others," Zamboni told me.

Robin Morris, a neuropsychologist at the Institute of Psychiatry at King's College London, would agree. Morris thinks anosognosia in Alzheimer's stems from a bigger problem than merely not remembering you have been diagnosed with the disease. We have, Morris argues, a special form of semantic memory that has to do with knowledge about ourselves—a self-representation system. This "personal database" is different from semantic knowledge about objects and facts about the world and other external things. "There is something particularly special about self-representation," Morris said when we met at his office in London. He hypothesized that in Alzheimer's, "people are not integrating new information into their self-representations."

According to Morris, this self-representation is essentially episodic memory that has somehow been turned into semantic memory about oneself—it's been semanticized, so to speak. Patient H. M. lends support to the idea that the essential meaning of our episodic memories is captured and stored in a semanticized form, separate from other episodic memories. When Suzanne Corkin asked H. M., "What is your favorite memory that you have of your mother?" he replied, "Well I, that she's just my mother." As Corkin found out, even though H. M. had memories of his childhood, "H.M. was unable to supply an episodic memory of his mother or his father—he could not narrate even one event that occurred at a specific time and place." Still, he had some sense of his pre-surgery self.

If our self-representation system is working well, then episodic memories are continually being converted to semantic memories, creating the gist of who we are. In Alzheimer's, the process seems to be disrupted. The brain's ability to continually update our self-representation is compromised, evidenced by the anosognosia, but more profoundly by an alteration of one's narrative self. The formation of the narrative slows down, or even stops. The patients reach back into the depths of their memories to a time when the keystones of their narrative arcs were formed, when their enduring identities were forged. In what can only be called a cruel twist of fate, this aspect of our sense of self that lets us travel back and forth in time, remembering or creating identity—what philosophers call autonoetic consciousness—is among the last of our cognitive abilities to mature during childhood, but one of the first to begin crumbling under the assault of Alzheimer's.

● ● ○

The day I met and talked with Allan, he was surprisingly present and charming. Michaele was glad he was so—it was that time of day, she said, early afternoon, a small window of awareness before he would lapse into a state where he would be difficult, even unreachable.

"Is there anything about Alzheimer's that worries you going forward?" I asked Allan.

"No, I think I have given up forward," he said. "I'm seventy, seventy-one." (He was eighty-one when we spoke, seventy when he was diagnosed.) "It's been a good life. Things could be much worse. Solved problems of the world, have two children, now two grandchildren. That's good. Saw the world when I was in the Air Force." It was when he was stationed in Germany with the US Air Force that he saw the destruction wrought by the Second World War. He was deeply moved by a visit to the site of a concentration camp in Dachau, which cemented his evolving view that the world would be better off being run by gypsies and ballet dancers—his shorthand for "people with little means and those with a love of the arts," said Michaele—than by dictators and politicians.

Michaele had pointed out—and I noticed it too—that Allan kept going back to his strongest memories: from his time spent in the Air Force to becoming a teacher of philosophy. There's a reason why he was doing this. It's a time when Allan was cementing his identity. "We think you incorporate those memories more powerfully into your sense of self, and you form richer, more enduring representations," Robin Morris told me. "Those basic building blocks—the essential concepts which define who you are—don't change over your lifetime, or they change more subtly."

End-stage Alzheimer's disease will get to those too, but for the time being, Allan had reserves that allowed him to remember back to when he was eighteen, when he got kicked out of school three times for smoking cigarettes. The school counselor suggested he join the Air Force, and Allan did. He was sent to a base near Munich, Germany, where he learned to be an airplane mechanic. He came back to San Francisco at the age of twenty-two and began working for United Airlines and attending community college. He wanted to become a radio broadcaster, but one of his teachers said he didn't have the voice to become a broadcaster, and encouraged him to take classes in philosophy. The advice turned out well—Allan loved philosophy, soon began teaching philosophy, and became a beloved teacher.

When I talked with Allan, he did not quite get the details or the sequence right (I could tell because Michaele had already filled me in).

For instance, when talking about the teacher who suggested he study philosophy, Allan said, "He was the one who said, 'Since you got kicked out . . . cigarettes, why don't you go join the Air Force.'" Actually, by the time he'd met the teacher, Allan had already been in the

Air Force, and the teacher had suggested that he take up philosophy classes.

If Michaele hadn't previously told me about Allan's past, I'd not have been able to temporally order his recollections. Toward the end of our meeting that day, Michaele left Allan and me alone to talk some more. I asked him again about his life. Here's a fragment of the conversation verbatim, when he talked of the Air Force again, after already having mentioned it a few times.

"We joined the Air Force here, and we continued to make up situations . . . you do. Everybody wanted to be an airplane flight, but then somebody said no, you have to have algebra. Some people said I don't have algebra. That's what I said. OK, we'll do something else. We got on the train in San Francisco, we went over up to Boston, got on a boat, a boat which is not much bigger than this room, and it was about twelve to fifteen people, and one of the finest things of all is that . . . many of those people . . . fallen sick. I'm one of the ones that didn't. We had to go out and puke. I had a picture of that once, yeah.

"Then we got off the train. About two and a half days to get to Munich. Munich is the center of Germany, that's where the place was. Making . . . Americans and Germans were together. We were there for . . . OK . . . two years . . . then we came back. I went to work for United Airlines, for years. Then everybody said, 'Why don't you go to college, because you are always reading books?' Which were detective books."

One vivid incident Allan mentioned several times during our conversation was the sight of farmers waving at the soldiers as their train sped through their fields. Allan would have been all of eighteen when he saw those farmers framed in the train window. But he couldn't correctly recollect whether he saw these farmers in Texas or in Germany. While Alzheimer's disease hadn't yet destroyed Allan's most vivid memories, it had scrambled his narrative.

● ● ○

In his office in London, Robin Morris pointed out two key changes that are happening in patients with Alzheimer's. One, as we saw earlier, is that they are not able to acquire new knowledge about themselves, and so are unable to update their narrative self. The other is that there are probably brain structures responsible for supporting our self that are under attack by Alzheimer's, and so the person is falling back on the most resilient parts of his or her narrative. These resilient notions of oneself form during late adolescence and early adulthood—much like the version of himself that Allan was recalling during his rambling recollections.

Even healthy people, when they are asked to recall life events, will remember more events from when they were between the ages of ten and thirty, compared with their recall of events from before and after this time. Psychologists have a name for this: the reminiscence bump.

This bump has a significant influence on our self. Martin Conway, a psychologist at City University London, has worked extensively on memory and the self. Conway envisages an individual as having a hierarchy of goals. This hierarchy is broken up into smaller and smaller subgoals, becoming more and more specific the smaller the scope of the subgoal. For instance, you might have as a goal the idea of becoming an athlete—and smaller and smaller subgoals would involve more and more specific targets, all the way down to running five kilometers today. Conway defines a notion of the "working self" based on this hierarchy of goals, and the purpose of the working self is to reconcile a specific goal (becoming an athlete) to the current state (sitting on the sofa, say) and ensure that the discrepancy between the two states is minimal (by making you get off that sofa and run). In other words, the working self regulates behavior.

Besides this working self, Conway identifies what he calls the conceptual self, an aspect of our self that contains notions of who we are, based on our interactions with others, including family, friends, society, and the broader culture.

The job of the working self, in Conway's model, is to regulate behavior and help in the formation and construction of memories that are consistent with the conceptual self and its goals. Consistency does not mean accuracy. For example, in the short term, it's important for me to remember whether I turned off the gas in the kitchen. Such short-term memory has to correspond to reality with high fidelity; otherwise there is a price to pay. However, it's impossible for the brain (at least in most of us who do not have photographic recall of every incident) to maintain such records ad infinitum. So, long-term memory is less constrained by fidelity and is more concerned with the need for "coherence"—that is, whatever enters long-term memory should not be dissonant with our conceptual self and goals, which are themselves influenced by existing long-term memories. As Conway puts it, the autobiographical knowledge in long-term memory "constrains what the self is, has been, and can be," while the working self dictates what goes into long-term autobiographical memory and the ease with which it can be accessed. Stories influence who we are, what we do, what we can be: certain beginnings require certain endings; stories can become our reality.

According to Conway, the neural processes that implement the working self also ensure that long-term memories that are coherent with our goals and self-knowledge are more easily accessible than those that are not. Crucially, memories of those experiences that were highly significant in meeting the goals of one's life seem to remain more strongly associated with the self and its history.

This brings us back to the reminiscence bump. "There is a critical period in late adolescence and early adulthood where you are defining your self-beliefs and self-concepts," Robin Morris told me. We form the core of our narrative self during this time.

So, the narrative self is influenced by highly significant events in one's life, and this self—or the memories associated with these events—then influences what you do next, and so dictates how your narrative grows. The self's need for coherence is paramount.

In patients with Alzheimer's, this narrative self is being disrupted on multiple fronts. To start with, the ability to form new episodic memories is impaired. Also, the incorporation of these memories into one's narrative in the form of gist or semanticized memories is failing too. Daniel Mograbi, Morris's PhD student from Rio de Janeiro, calls this the petrified self: the story one is able to tell about oneself once Alzheimer's takes hold is stalled. When the narrative self is functioning normally, episodes in one's life line up to tell a story. In its early stages, Alzheimer's prevents the narrative arc from growing any further, limiting it to whatever it was at the time of disease onset. Alzheimer's continues to hack away at the narrative until all one is left with is a set of disconnected episodes. Eventually, even those are gone.

The term "petrified self" did not go down too well among some of their colleagues. "It suggests the person with Alzheimer's is dead or ossified," said Morris. "I have a lot of sympathy for that, since that is not what we meant. We should be careful how we conceptualize people. On the other hand, you can't limit science by political correctness. You can't hide inconvenient truths."

And the truth is that as the narrative self petrifies first and then begins to deteriorate, the person with Alzheimer's reverts back to the critical narrative self, to memories that were formed at a time when the self was being defined most strongly, when its essence was being etched deeply in the body and brain. Alzheimer's, however, ultimately affects even the critical narrative self. Despite Allan's ability to reminisce about his adolescence and early adulthood, Michaele would notice long periods when he would just "disappear, disappear." She'd look into his eyes and find them empty, vacant. All caregivers of Alzheimer's patients would relate to Michaele's experience. "There was just hardly [anyone] there anymore," she told me.

But is that something caregivers are inferring, or is the person with Alzheimer's really unaware? Morris argues that the onus is on science to show that patients are not aware, that they have no self.

● ● ○

Pia Kontos is not comfortable with claims that Alzheimer's patients ultimately have no self. She argues that even in the face of severe cognitive decline evident in Alzheimer's patients, a form of selfhood persists, a precognitive, prereflective selfhood that's embedded in the body. She takes her inspiration from French philosopher Maurice Merleau-Ponty and French sociologist Pierre Bourdieu. "Bourdieu and Merleau-Ponty help [us] think about what the body brings to our engagement with the world that doesn't rely on cognition," she told me.

She has seen examples of such "embodied selfhood" in her research in long-term care settings with people with Alzheimer's disease. One particular observation—of an elderly male resident who was severely cognitively impaired and spoke only in single words, often nonsensical—left a deep impression on her. One day, on Simchat

Torah, a Jewish high holiday to celebrate the Torah, the residents went to the synagogue in the long-term care home. The old man stood in line, waiting to be called to the bimah (pulpit) to sing the prayer. "I saw this gentleman get up in the lineup, and I remember my whole body clutched," said Kontos. "This is going to be a disaster, I thought, because he can't put two words together."

What followed stunned her. When his name was called out, the man confidently walked up to the bimah and recited the prayer with utter proficiency. One could argue that there was some cognition still intact in him that allowed him to do so. But Kontos thinks otherwise.

"The way that I have analyzed it is that there was an orchestration of an event there. There was the touch of the Torah, the presence of the rabbi, the presence of all the congregants, and that elicited in him what Bourdieu has termed 'habitus,' but I term 'embodied selfhood,' and it enabled him to perform in that moment," Kontos said. "If you took that gentleman to his room and asked him to recite the prayer, he couldn't do it."

Embodied selfhood is "the idea that bodily habits, gestures, and actions support and convey humanness and individuality." Merleau-Ponty argued that we are all born with a primordial body that is capable of engaging with the world. "Nothing human is altogether incorporeal," he wrote. He took as an example the skill of touch-typing. If you are a capable touch typist, then typing for you is an activity that does not require you to think about the location of keys on a keyboard. "Knowledge of typing," Merleau-Ponty argues, "is in the hands and manifests itself only when bodily effort is made and cannot be articulated in detachment from that effort."

Bourdieu extended the role of the body beyond its primordial capacity: the body, he said, incorporates our social and cultural habits.

This is what gave rise to the name "habitus." "Habitus comprises dispositions and forms of know-how, which function below the threshold of cognition and are enacted at a prereflective level," writes Kontos, where a disposition is "a way of being, a habitual state . . . a tendency, propensity, or inclination," according to Bourdieu.

Kontos combines Merleau-Ponty's primordial body with Bourdieu's habitus to come to her notion of embodied selfhood. "We all have embodied selfhood; you have it, I have it. It's just that when our cognition is intact, it goes unnoticed; it is sort of in the background. But when we have cognitive impairment, it comes to the foreground," she told me. "This prereflective ability to engage with the world becomes even more important in the face of cognitive impairment because it becomes the primary means of engaging with the world."

Embodied selfhood is blurring the distinction between body and mind; it gives the body its due in making us who we are. Thanks to Descartes, Western neuroscience had elevated the mind, demoting the body to being a mere container, and while neuroscience has slowly distanced itself from Descartes and done away with a stark split between body and mind, the legacy of centuries of such thinking still leads us astray when we attribute to Alzheimer's patients a total loss of self. "Because of Cartesianism, and this constant devaluation of the body, what happens is that when we lose cognitive ability, people jump very quickly to the assumption that there is no self," said Kontos. "But there is still this fundamental dimension of our existence that persists." If we drop Descartes's legacy entirely, and stop distinguishing between body and mind, a new perspective on the self begins to emerge.

So, embodied selfhood involves the brain *and* the body, but in a way that does not necessarily involve cognition. The brain is broadly divided into three physical regions: the cerebral cortex, the cerebellum, and the brain stem. The cerebellum plays a significant role in procedural memory and in coordinating how our bodies move, and it survives more or less intact until the very late stages of Alzheimer's disease. So, even as the cerebral cortex atrophies and cognition declines, some parts of the brain-body complex continue to store and play out aspects of our selfhood.

Another person living with Alzheimer's disease drove home this message for Kontos. She was an elderly woman, with cognitive impairment so severe that she could not speak, or even dress or feed herself, and was confined to her wheelchair. She was also incontinent. When the nursing staff would wheel her to the dining room and put a bib on her (institutional policy to prevent patients from soiling themselves), she'd struggle to reach under the bib and pull out a string of pearls that she wore, and rest it on top of the bib, where they could be seen. "She'd never begin her meal until she did that," Kontos told me. "She emerges from the depths of dementia with a very strong presence. If that's not self-expression, I don't know what is."

But the making of a complex narrative selfhood (cognitive or embodied) might involve something even more fundamental: the ability to simply be the subject of an experience. When I met Allan, it was clear that despite his incoherent storytelling, he still was someone who was experiencing his own scrambled narrative. It's possible that during the very late stages of Alzheimer's disease, when one's narrative self is completely destroyed, all that is left is the self-as-subject, experiencing those aspects of the self that exist even before a narrative forms. One could argue that the self at its most fundamental is the self-as-subject, and it's not one's narrative. Who or what is this self-as-subject? Sadly, those in the grips of the disease cannot communicate what it's like to be without a narrative—and it'd be too cruel to ask.

We have to turn to clues from elsewhere to understand the basis of this subjectivity. For example, for touch-typing to become an embodied ability and a part of my extended narrative self, do I need to feel the touch of a key at my fingertips and know that *I touched the key*, as opposed to feeling as if someone else was doing it? Or, for that matter, don't I need to feel that my fingers are *my own*? These might seem like outrageous questions, but the next chapter will show that something we take for granted—ownership of body parts—can be disrupted, in experiments and pathologically. When it's the latter, the consequences can be unimaginably dire.

● ● ○

Billy Joel's "It's Still Rock and Roll to Me" is playing on the car radio as I pull into the parking lot of the assisted-living facility to see Clare's father. "*What's the matter with the clothes I'm wearing / Can't you tell that your tie's too wide*?" The California afternoon sun is hot, accentuated by my car's barely functional air-conditioning. Clare is waiting outside the building for me. She punches a security code to enter, a precaution not to prevent outsiders from entering, but rather to keep the residents—mainly patients with Alzheimer's—from wandering out, which they tend to do. We walk down the corridors, past her dad's room (a sign wishing him a happy birthday when he turned ninety last month still hangs on the door). A couple of elderly ladies smile at us and one says, "Good morning." After a moment's pause she adds, "Or good afternoon, I don't know." I can't tell whether she's indulging in some inside humor or not. Either way, it imbues the place with a tender resilience.

We go to see Clare's father in a large hall. It's a scene I have seen only in movies. About twenty men and women, all elderly, are sitting, some slumped over, some relatively alert. A television set is playing a movie, loudly. It's a recent Michael Caine movie (*Last Love*, I find out later). Clare points to her dad—he's sitting in his own chair, which Clare's mother had brought over so that he'd be more comfortable than in the standard-issue chairs. He is asleep. Clare walks over and gently nudges him. "Dad, Dad," she says. He wakes up perturbed, agitated. Clare reaches for his hands, but he angrily swats at her hand. She tries again to hold his hand, and he reaches out in a handshake gesture, only to twist her hand. She pulls away. He's clearly upset at being woken up. We leave him for the moment and go to his room.

Clare has a key to enter her dad's room—the rooms are locked because otherwise the patients would go around opening doors and entering. The room is simple and sparse. Framed pictures hang on the walls, reminders of Clare's father's life. There's one of him looking handsome at the wheel of a sailboat. There are many family photographs. On the table is a scrapbook made of colored construction paper, the kind a child would make. It's actually something Clare's sister has made for their dad—a simple storybook of some key moments in his life: a photograph of him when he was seventeen; Clare's dad and mother signing the marriage register in Europe; their fellow rowers holding up their oars to form an arch for the bride and groom to walk under as they come out of the church; the family at a beach in Morro Bay, California, after they had come to America—Clare and her sisters are little girls; their home in Minnesota, where Clare grew up; building a barbecue pit at home (one of the few times Clare's dad actually worked with his hands); Clare's dad on the cover of his company's magazine, pictured as a captain on a boat; a trip during a wedding anniversary, around when he turned seventy; and a photograph of Clare's dad from ten years ago. "There's been a tremendous decline since then," says Clare.

Clare's sister's attempt to jog her father's memory, to give him back his narrative, his coherent story, his self, with this scrapbook hasn't made much of a difference as far as Clare can see.

We go back to see Clare's dad. This time he lets Clare hold his hand briefly. He even squeezes her knuckles. Clare blows a kiss at him, and a few kisses later, he smiles and does the same. I turn to Clare and ask if that means he recognizes her. She says she doesn't know. He hasn't said a word. There's no way to tell. I too try to shake his hand; he doesn't respond at first, but then for a brief moment, he smiles and shakes my hand firmly. He then squeezes my knuckles too. There's no way to tell.

Or maybe there is. For Clare, the knuckle-squeezing takes her back to childhood, when her dad would do that to her playfully. It would make her wince. "'Ha-ha-ha, little joke,'" he'd say, Clare recalled. Could it be that somewhere in that body, Clare's dad still persists—a fragment of his self, a memory, a strong, strapping man still playing with his daughter?

● ● ○

About a month and a half after I met Allan, Michaele took him to check out a board-and-care home. Allan had been incontinent for days, suffering from severe diarrhea. Michaele had spent sleepless nights changing sheets and giving Allan numerous showers. Realizing that they needed help, she drove Allan to visit the care home, which was beautiful, with a nice backyard full of trees overlooking a park. Allan seemed to like it. As they were driving away from the home, Michaele said to him, "Do you think you'll be OK there?" To her surprise, he answered, "I think it's nice, it'll be nice."

He said it with such lucidity that Michaele was immediately guilt-stricken. "Oh, Allan, I feel terrible. I'm going to miss you so much. It's so hard for me to do this. But I know I can't keep going on," she told him.

"That's OK," he said. "We will always be connected no matter what happens."

"That blew me away," Michaele told me. "His ability to communicate with me so clearly that day was phenomenal. He got very quiet again. But I just felt so close to him that day."

Allan would spend just two weeks in the board-and-care home, and then he passed away.

I met Michaele a few weeks after Allan's death. We sat in the same living room where I had first met him. On a small table next to Allan's brown leather sofa, Michaele had set a small white vase full of fresh flowers from their garden, and placed a small clay tortoise atop a few of his favorite books. A candle of lavender-colored wax burned beside a framed photograph of a younger Michaele and Allan. On the sofa's high back, Michaele had carefully draped Allan's brown corduroy jacket.

3

THE MAN WHO DIDN'T WANT HIS LEG

IS THE FEELING THAT YOU OWN YOUR BODY AND ITS VARIOUS PARTS BASED ON REALITY?

The leg suddenly assumed an eerie character—or more precisely, if less evocatively, lost all its character—and became a foreign, inconceivable *thing*, which I looked at, and touched, without any sense whatever of recognition or relation. . . . I gazed at it, and felt, I don't know you, you're not part of me.

—Oliver Sacks

Theoretically you can have a phantom of almost any part of the body, except of course the brain; you can't have a phantom brain, by definition, because that's where we think it's all happening.

—V. S. Ramachandran

This wasn't the first time that David had tried to amputate his leg. When he was just out of college, he had tried to do it using a tourniquet fashioned out of an old sock and strong baling twine. David locked himself in his bedroom at his parents' house, his bound leg propped up against the wall to prevent blood from flowing into it. After two hours the pain was unbearable, and fear sapped his will. Undoing a tourniquet that has starved a limb of blood can be fatal: injured muscles downstream of the blockage can flood the body with toxins, causing the kidneys to fail. Even so, David released the tourniquet himself; it was just as well that he hadn't mastered the art of tying one.

Failure did not lessen David's desire to be rid of the leg. It began to consume him, to dominate his awareness. The leg was always there as a foreign body, an impostor, an intrusion. He spent every waking moment imagining freedom from the leg. He'd stand on his "good" leg, trying not to put any weight on the bad one. At home, he'd hop around. While sitting, he'd often push the leg to one side. The leg just wasn't his. He began to blame it for keeping him single; but living alone in a small suburban town house, afraid to socialize and struggling to form relationships, David was unwilling to let anyone know of his singular fixation.

David is not his real name. He wouldn't discuss his condition without the protection of anonymity. After he'd agreed to talk face-to-face, we met in the waiting area of a nondescript restaurant, in a nondescript mall just outside one of America's largest cities. A handsome man, David resembles a certain edgy movie star whose name, he feared, might identify him to his coworkers if I revealed it. He had kept his secret well hidden: I was only the second individual whom he had confided to in person about his leg.

The cheerful guitar music in the restaurant lobby clashed with David's mood. He choked up as he recounted his depression. I'd heard his voice cracking when we'd spoken earlier on the phone, but watching this grown man so full of emotion was difficult. The restaurant's buzzer went off. Our table inside was ready, but David didn't want to go in. Even though his voice was shaking, he wanted to keep talking.

"It got to the point where I'd come into my house and just cry," he had told me earlier over the phone. "I'd be looking at other people and seeing that they already have their lives going good for them. And I'm stuck here, all miserable. I'm being held back by this strange obsession. The logic going through my head was that I need to take care of this now, because if I wait any longer, there is not much chance of a life for me."

It took some time for David to open up. Early on, when we were just getting to know each other, he was shy and polite, confessing that he wasn't very good at talking about himself. He had avoided seeking professional psychiatric help, afraid that doing so would somehow endanger his employment. And yet he knew that he was slipping into a dark place. He began associating his house with the feeling of being alone and depressed. Soon he came home only to sleep; he couldn't be in the house during the day without breaking into tears.

One night about a year before I met him, when he could bear it no longer, David called his best friend. There was something he had been wanting to reveal, David told him. His friend's response was empathetic—exactly what David needed. Even as David was speaking, his friend began searching online for material. "He told me that there was something in my eyes the whole time I was growing up," David said. "It looked like I had pain in my eyes, like there was something I wasn't telling him."

Once David opened up, he discovered that he was not alone. He found a community on the Internet of others who were also desperate to excise some part of their body—usually a limb, sometimes two. These people were suffering from what is commonly called body integrity identity disorder (BIID). The scientific community is debating whether BIID is the correct name for the condition. They have also suggested xenomelia, from the Greek for "foreign" and "limb," but I'll stick to BIID in this chapter.

The online community has been a blessing to those who suffer from BIID, and through it many discover that their malaise has an official name. With a handful of websites and a few thousand members, the community even has its internal subdivisions: "devotees" are fascinated by or attracted to amputees, often sexually, but don't want amputations themselves; "wannabes" strongly desire an amputation of their own. A further delineation, "need-to-be," describes someone whose desire for amputation is particularly fierce.

It was a wannabe who told David about a former BIID patient who had been connecting other wannabes to a surgeon in Asia. For a fee, this doctor would perform off-the-books amputations. David contacted this gatekeeper on Facebook, but more than a month passed without a reply. As his hopes of surgery began to fade, David's depression deepened. The leg intruded more insistently into his thoughts. He decided to try again to get rid of it himself.

Rather than resorting to a tourniquet, this time he settled for dry ice, one of the preferred methods of self-amputation among the BIID community. The idea is to freeze the offending limb and damage it to the point that doctors have no choice but to amputate. David drove over to his local Walmart and bought two large trash cans. The plan was brutal but simple. First, he would submerge the leg in a can full of cold water to numb it. Then he would pack it in a can full of dry ice until it was injured beyond repair.

He bought rolls of bandages, but he couldn't find the dry ice or the prescription painkillers he needed if he was going to keep the leg in dry ice for the required eight hours. David went home despondent, with just two trash cans and bandages, preparing himself mentally to go out the next day to find the other ingredients. The painkillers were essential; he knew that without them he would never succeed. Then, before going to bed that night, he checked his computer.

There it was: a message. The gatekeeper wanted to talk.

● ● ○

We are only just beginning to understand BIID. It hasn't helped that the medical establishment generally dismisses the condition as a perversion. Yet there is evidence that it has existed for hundreds of years. In a recent paper, Peter Brugger, the head of neuropsychology at University Hospital Zurich, Switzerland, cited the case of an Englishman who went to France in the late eighteenth century and asked a surgeon to amputate his leg. When the surgeon refused, the Englishman held him up at gunpoint, forcing him to perform the operation. After returning home, he sent the surgeon 250 guineas and a letter of thanks, in which he wrote that his leg had been "an invincible obstacle" to his happiness.

The first modern account of the condition dates from 1977, when *The Journal of Sex Research* published a paper on "apotemnophilia"—the desire to be an amputee. The paper categorized the desire for amputation as a paraphilia, a catchall term for deviant sexual desires. Although it's true that most people who desire such amputations are sexually attracted to amputees, the term "paraphilia" has long been a convenient label for misunderstandings. After all, at one time homosexuality was also labeled as paraphilia.

One of the coauthors of the 1977 paper was Gregg Furth, who eventually became a practicing psychologist in New York. Furth himself suffered from the condition and, over time, became a major figure in the BIID underground. He wanted to help people deal with their problem, but medical treatment was always controversial—often for good reason. In 1998, Furth introduced a friend to an unlicensed surgeon who agreed to amputate the friend's leg in a clinic in Tijuana, Mexico. The patient died of gangrene and the surgeon was sent to prison. Around the same time, a Scottish surgeon named Robert Smith, who practiced at the Falkirk and District Royal Infirmary, briefly held out legal hope for BIID sufferers by openly performing voluntary amputations, but a media frenzy in 2000 led British authorities to forbid such procedures. The Smith affair fueled a series of articles about the condition—some suggesting that merely identifying and defining such a condition could cause it to spread, a form of cultural contagion.

Undeterred, Furth found a surgeon in Asia who was willing to perform amputations for about $6,000. But instead of getting the surgery himself, he began acting as a go-between, putting sufferers in touch with the surgeon. He also contacted Michael First, a clinical psychiatrist at Columbia University in New York. Intrigued, First embarked on a survey of fifty-two patients. What he found was instructive. The patients all seemed to be obsessed by the thought of a body that was different in some way from the one they possessed. There seemed to be a mismatch between their internal sense of their own bodies and their actual physical bodies. First, who would later lobby to have BIID more widely recognized, became convinced that he was looking at a disorder of identity, of the sense of self.

"The name that was originally proposed, 'apotemnophilia,' was clearly a problem," he told me. "We wanted a word that was parallel to gender identity disorder. GID has built into the name a concept that there is a function called gender identity, which is your sense of being male or female, which has gone wrong. So, what would be a parallel notion? Body integrity identity disorder hypothesizes that a normal function, which is your comfort in how your body fits together, has gone wrong."

In June 2003, First presented his findings at a meeting in New York. Robert Smith, Furth, and many BIID sufferers attended the meeting. One of them was David's gatekeeper, whom I'll call Patrick.

Without much warning, Furth walked up to Patrick and his wife with a startling proposition. "We are standing there eating our sandwich, and he says to me, 'Would you be interested in a surgical option?'" Patrick had felt the pressure of BIID for most of his life. He didn't have any reservations. "Hell, yes. Yes, yes, yes, no question about it." To this day, Patrick doesn't know why Furth singled him out. Patrick is not a religious man, but he felt a higher power was giving him his due.

The next evening, Patrick and his wife went over to Furth's apartment for an evaluation. Furth grilled Patrick to make sure he was for real. Was Patrick's desire due to BIID or a sexual fetish? How did it affect his life? For two hours the questions flowed. Patrick answered them, scared that he'd "flunk the evaluation." He didn't, and Furth agreed to make the recommendation. That was where it all began. Ten months later, he had the surgery he craved. And less than a year after that, Patrick had become the gatekeeper himself.

● ● ○

Sitting at home in a small, somewhat rural American town not too far from the ocean, Patrick recalled the day his wife found out about his obsession. It was during the mid-'90s. As with almost all BIID sufferers, Patrick was fascinated with amputees, so he began downloading pictures of them off the Internet and printing them out. One day his wife was sitting in front of their computer, while Patrick sat in a wingback chair. She noticed a pile of printouts. They were images of men, but "completely clothed, no nudes or anything like that." It was an awkward moment. "She was thinking that maybe I was gay," Patrick recalled. "I must have been crimson." Patrick asked her to take a closer look. She did, and soon realized that the men were all amputees.

Patrick told his wife that he had felt odd about his leg since he was four years old, a feeling that eventually grew into an all-consuming desire to be rid of it. It was a shock: they had been married for decades, and the revelation that he had been hiding something from her for all those years was hard to take. But his confession also brought relief. For more than four decades Patrick had suffered alone. Growing up in small-town America, with conservative parents, in an era when "people didn't believe in going and seeing mental-health professionals," Patrick was mystified by what he felt. By the early '60s, as a teenager, his obsession with amputees and amputations took him to a library in the nearby state capital, where he hoped to find books on the subject. To his surprise, most of the pictures of amputees had been cut out and stolen. At that moment he realized that he wasn't the only person who was consumed by this strange obsession.

"There had to be somebody else out there," Patrick told me, "but how could I find out who?"

As time went on, Patrick struggled with his thoughts about his leg: "How can I get rid of it? What can I do? How can I do it? I don't want to die in the process." Seeing a picture of an amputee, or worse, seeing an amputee on the streets, would ratchet up his emotions. "It would just drive me nuts," he told me. "That could last for several days. All I could think about was how I could get rid of my leg." His anxiety led him to make deals with God and pacts with the devil: "Take my leg, save somebody else's," he implored. Yet through it all, for the first four and a half decades of his life, he told no one. The loneliness was almost too much to bear.

Less than a year before his wife's discovery, he had stumbled upon an anonymous classified ad in a local city newsletter. The person who placed it admitted a desire to amputate a limb; he was a wannabe. Patrick wrote to the PO box that was listed and began a correspondence with the man. Eventually they met, and the wannabe told him about others who were seeking amputations. It was deliverance. "Oh my God, I'm not alone with this anymore," Patrick recalled thinking. "I'm not nuts."

Yet finding others who shared his condition did not lessen his need. If anything, Patrick's desperation grew. He considered a DIY amputation. He had heard of people who had lain down on train tracks and let a train run over their limbs, or who had blown their legs off with a shotgun. "The trouble with a train is that if the train is moving at a good clip, you can kill yourself very easily, because it can pick you up and spit you out," he said. "I really didn't want to die in the process and not find out what it was like to live with one leg."

Another wannabe who had done a DIY amputation suggested Patrick practice first, so Patrick decided to get rid of part of his finger as a prelude to amputating his leg. With a pen and a rubber band, he made a tourniquet for one of his fingers and stuck it into a thermal cup full of ice and alcohol. After part of the finger became numb and Patrick was unable to bend it, he took a hammer and chisel and chopped off the bit above the first knuckle. He even smashed the detached digit. "So they couldn't reattach it even if they wanted to," Patrick told me.

Crushing the amputated digit also aided in the cover-up: hospital staff were told that a heavy object had fallen on the finger. When a doctor injected his injured finger with a painkiller, Patrick pretended that the needle hurt. His finger was still too numb to feel any pain.

● ● ○

It was about a decade ago that Patrick finally made it to Asia to see the surgeon Greg Furth introduced him to. He was admitted to the hospital on a Friday evening and had to wait until Saturday evening to be wheeled into surgery. "The single longest day of my life," he told me. He awoke from his anesthesia the next day. "I looked down and couldn't believe it. It was finally gone," he said. "I was ecstatic." His only regret in the ten years since his amputation is that he didn't get it when he was younger. "I wouldn't want my leg back for all the money in the world, that's how happy I am."

This comfort with his condition is reflected at home. Just before the surgery, his children gave him a Ken doll, which he keeps in a plastic box stuffed with scrapbooks of photographs of amputees that he collected in his younger days. The doll wears a pair of red shorts; one of its legs ends above the knee, in a stump wrapped with a white gauze bandage. In Patrick's house, I saw a decorative skeleton hanging off a chandelier and didn't think much of it. "Look more closely," he urged. Only then did I notice that it, like Patrick, was missing part of a leg and part of one finger. Then there was a statue of Michelangelo's *David* on the mantelpiece. It, too, was missing part of a leg. The family had acknowledged Patrick's suffering and was celebrating his freedom from BIID. Patrick now seemed genuinely comfortable with his body.

This feeling of relief and release is a sentiment expressed by just about every BIID amputee who has been studied by scientists. That evidence ought to allay at least one fear that ethicists have expressed about BIID—that once you amputate a healthy limb, the patients will eventually come back for more. In nearly all accounts, they don't, unless from the very beginning their BIID involves multiple limbs.

Furth, for his part, was diagnosed with cancer and died in 2005 without ever getting his own amputation. When he vetted Patrick for surgery, Patrick told him that after his amputation he would try to help the others he knew were out there. Nearing death, Furth called Patrick. Would he take over the gatekeeping duties for the surgeon in Asia? Patrick agreed to do so, and for nine years he has acted as the go-between for BIID sufferers. One way or another, they eventually find him. And just before he could use dry ice on his leg, David found him too.

● ● ○

A year or so before Patrick's operation, a psychologist asked him if he would take a pill to make his BIID go away, should such a treatment exist. It took a moment for him to reflect and answer: maybe when he had been a lot younger, but not anymore. "This has become the core of who and what I am," he said.

This is who I am. Everyone with BIID that I interviewed or heard about uses some variation on those words to describe their condition. When they envision themselves whole and complete, that image does not include parts of their limbs. "It seems like my body stops mid-thigh of my right leg," Furth told the makers of a 2000 BBC documentary, *Complete Obsession.* "The rest is not me."

In the same film, the Scottish surgeon Robert Smith tells an interviewer, "I have become convinced over the years that there is a small group of patients who genuinely feel that their body is incomplete with their normal complement of four limbs."

It's difficult for most of us to relate to this. Your sense of self, like mine, is probably tied to a body that has its entire complement of limbs. I can't bear the thought of someone taking a scalpel to my thigh. It's *my* thigh. I take that sense of ownership for granted. This isn't the case for BIID sufferers, and it wasn't the case for David. When I asked him to describe how his leg felt, he said, "It feels like my soul doesn't extend into it."

Neuroscience has shown us over the past decade or so that this sense of ownership over our body parts is strangely malleable, even among normal healthy people. In 1998, cognitive scientists at Carnegie Mellon University in Pittsburgh performed an ingenious experiment. They sat subjects down at a table and asked them to rest their left hands on it. A rubber hand was placed next to the real hand. The researchers put a screen between the two, so that the subjects could see only the rubber hand but not the real hand. The researchers then used two small paintbrushes to stroke both the real hand and the rubber hand at the same time. When questioned later, the subjects said that they eventually felt the brush not on their real hand but on the rubber hand—even though they were fully cognizant at all times that their real hand was being brushed. More significantly, many said they felt as if the rubber hand was their own.

The rubber-hand illusion illustrates how the way we experience our body parts is a dynamic process, one that involves constant integration of various senses. Visual and tactile information, along with sensations from joints, tendons, and muscles that give us an internal sense of the relative position of our body parts (neuroscientists call this sense proprioception), are combined to give us a sense of ownership of our bodies. This feeling is a crucial component of our sense of self. It's only when the process that creates this sense of ownership goes awry—for example when the brain receives conflicting sense information, as in the rubber-hand illusion—that we notice something is amiss.

It's likely that the brain has different mechanisms to create a sense of ownership. For instance, as we'll see in the next chapter, the brain creates a sense of being the initiator of one's thoughts and actions—the feeling that *you* performed an action when you picked up a bottle, or that *you* thought something and it felt like *your* thought, and not someone else's. This so-called sense of agency is key to owning your actions and your thoughts (when it goes wrong, the consequences can be debilitating, including psychotic delusions and schizophrenia).

So, if we can feel as if we own something as inanimate as a rubber hand, can we own something that doesn't exist? Seemingly, yes. Patients who have lost a limb can sometimes sense its presence, often immediately after surgery and at times even years after the amputation. In 1871, an American physician named Silas Weir Mitchell coined the phrase "phantom limb" for such a sensation. Some patients can even feel pain in their phantom limbs. By the early 1990s, thanks mainly to some pioneering work by neuroscientist V. S. Ramachandran of the University of California, San Diego, it was established that phantom limbs were an artifact of body representation in the brain gone wrong.

The idea that our brain creates maps or representations of the body emerged in the 1930s, when Canadian neurosurgeon Wilder Penfield probed the brains of conscious patients who were undergoing neurosurgery for severe epilepsy. He found that each part of the body's outer surface has its counterpart on the surface of the brain's cortex: the more sensitive the body part—say, hands and fingers, or the face—

the larger the brain area devoted to it. As it turns out, the brain maps far more than just the body's outer surface. According to neuroscientists, the brain creates maps for everything we perceive, from our bodies (both the external surface and the interior tissues) to attributes of the external world. These maps compose the objects of consciousness.

The presence of such maps can explain phantom limbs. Though patients have lost a limb, the cortical maps sometimes remain—intact, fragmented, or modified—and they can lead to the perception of a limb, along with its potential to feel pain. Even people born without limbs can experience phantom arms or legs. In 2000, Peter Brugger wrote about a forty-four-year-old highly educated woman, born without forearms and legs, who nonetheless had experienced them as phantom limbs for as long as she could remember. Using fMRI and transcranial magnetic stimulation (TMS), Brugger's team verified her subjective experience of phantom limbs and showed that body parts that were absent from birth could still be represented in sensory and motor cortices. "These phantoms of congenitally absent limbs are *animation without incarnation*," Brugger told me. "Nothing had ever turned into flesh and bones." The brain had the maps for the missing body parts even though the actual limbs had failed to develop.

When confronted with BIID, Brugger saw parallels to what the forty-four-year-old woman experienced. "There must be the converse, which is an *incarnation without animation*," he said. "And this is BIID." The body had developed fully, but somehow its representation in the brain was incomplete. The maps for a part of a limb or limbs were compromised.

Recent studies have borne out this idea. Neuroscientists are particularly interested in the right superior parietal lobule (SPL), a brain region thought to be vital to the construction of body maps. Brugger's team has found that this area is thinner in those with BIID, and others have shown that it may be functioning differently in those with the condition. In 2008, Paul McGeoch and V. S. Ramachandran mapped the activity in the brains of four BIID patients and control subjects. The researchers tapped the feet of the control subjects and watched the SPL light up. But the BIID patients were different: the right SPL showed reduced activity when the disowned foot was tapped, only lighting up normally when the tap was on the other foot.

"What we argue is that in these people something has gone wrong in the development, either congenitally or in the early development, of this part of the brain," McGeoch told me. "This limb is not adequately represented. They find themselves in a state of conflict, a state of mismatch that they can see and feel."

There are almost certainly other parts of the brain involved. Recently, scientists reviewed a number of "body-ownership" experiments, including the rubber-hand illusion, and identified a network of brain regions that integrate sensations from the body and its immediate surroundings and sensations related to the movement of our body parts. The network includes a clutch of areas, from cortical regions responsible for motor control and the sense of touch all the way to the brain stem. This network, they suggest, is responsible for what they call the "body matrix"—a sense of our physical body and the immediate space around it. Because the network helps maintain the internal physiological balance of the body, it reacts to anything that threatens the body's integrity and stability. Intriguingly, the physical differences in the brains of BIID patients that Brugger identified include changes in nearly all the parts of this network. Could BIID result from alterations to this body-matrix network? Brugger's team thinks so.

It's crucial to emphasize that these findings are *correlations*—they don't prove that the neural anomalies are the *cause* of BIID. It's a caveat to keep in mind throughout this book. There is a tendency within neuroscience toward neurobiological reductionism, especially in the study of disorders, by viewing the brain-mind relationship as a one-way street, with the brain influencing mental activity, and not the other way around. fMRI or PET scans usually tell us about the relative change in activity in specific brain regions in a person with some disorder when compared with healthy controls. But except in clear-cut cases of neurological damage, such scans give us correlations between brain activity and a person's condition; they don't definitively establish whether the observed anatomical and functional aberrations seen in brain scans came first and caused someone's condition (such as BIID) or whether ceaseless mental activity (thinking obsessively that "this leg is not mine," for example) led to the changes in the brain.

Then there is the question of how body states and body-matrix networks translate into a sense of self. And for BIID patients, how does a skewed body map lead to the desire for amputation?

Philosopher Thomas Metzinger provides an insight into why someone with BIID might disown a body part, and it has to do with his ideas about the self. "'Owning' your body, its sensations, and its various parts is fundamental to the feeling of being someone," Metzinger wrote in his book *The Ego Tunnel.* In his theory, the brain creates a model, a representation of the environment in which the body exists. Embedded within this model of the world is a model of the self: a representation of the organism itself, which is used to "regulate its interaction with the environment" and to maintain the organism in an optimal state of functioning.

That the brain must create such models follows from a classic 1970 paper that showed mathematically that "any regulator . . . must model what it regulates." So, if the brain is trying to regulate the body, it must model the body, and this is the self-model.

Crucially, only a subset of this self-model enters conscious awareness. This is what Metzinger calls the phenomenal self-model (PSM). The contents of this model are what we are conscious of, including bodily sensations, emotions, and thoughts. Put another way, the content of the PSM is our ego, our identity as subjectively experienced. At any given moment, there might be body states that are part of the self-model but not part of the PSM, in which case we would not be conscious or subjectively aware of those body states. And the contents of these models, whether of the world, the self, or the phenomenal self, are constantly changing. Also, what separates the contents of the world-model from the contents of the PSM is the property of *mineness*: objects in the world-model don't feel like mine, while those in the PSM, whatever they are, by definition feel like they belong to me.

If Metzinger is right, then before the rubber-hand illusion sets in, the lifeless hand that one is seeing is part of one's world-model, but not of the PSM. So, it lacks a sense of belonging to *me*. We come under the grip of the illusion because the experiment modifies our PSM: our brains replace the representation of the real hand with a representation of the rubber hand, which is now embedded in our phenomenal self-model. Since anything in the PSM has the subjective property of *mineness*, we feel as if the rubber hand belongs to us. In BIID, it's likely that a limb or some other body part is misrepresented or underrepresented in the PSM. Lacking the property of mineness, it is disowned (it's intriguing to think that Cotard's syndrome could also be due to a messed-up phenomenal self-model).

Metzinger's ideas give us a clue to why someone with BIID might want to amputate a limb that doesn't feel like it belongs. My self—as defined by the content of the PSM—is not just my subjective identity; it is also the basis for the boundary between what's mine and everything else, between me and not-me. "It's a tool and a weapon," said Metzinger, when we spoke on the telephone. "It's something that evolved to constantly preserve and sustain and defend the integrity of the overall organism, and that includes drawing a line between me and not-me on very many different functional levels. If there is a misrepresentation in the brain that tells you this is not your limb, it follows that this will be a permanently alarming situation."

McGeoch, Ramachandran, and their colleagues showed this in a simple and elegant experiment. They studied two people who wanted voluntary amputations: a twenty-nine-year-old man who desired an amputation below his right knee; and a sixty-three-year-old man who wanted amputations below his left knee and below his right thigh. One of the curious things about BIID is that most sufferers can precisely delineate their limb into the part that feels their own and the part that doesn't. And this separation is stable over time (suggesting that the condition is neurological rather than psychological, according to Ramachandran's team). In their study, the researchers recorded skin conductance response (SCR) using electrodes attached to the subjects' hands, when they were pricked with a pin either below or above the "desired line of amputation."

SCR cannot be willfully controlled. Most people, when they are touched or when they hear a noise or when they perceive emotionally salient stimuli, will show an increased SCR. Ramachandran's study showed that when their BIID subjects were pricked on the part of the limb that they felt was foreign, the SCR was two to three times greater than when the pinprick was on the normal part of the limb. One interpretation of this data is that the pinprick on the part they wanted amputated was felt as more threatening.

Brugger's team also found something similar. When BIID patients were tapped simultaneously on both the foreign-feeling and normal parts of their limbs, they reported feeling the taps on the rejected parts earlier: their brains were prioritizing these tactile stimuli.

Both studies suggest a hyperawareness of the disowned parts of limbs. It's as if the brain of BIID sufferers is paying extra attention to body parts that feel alien. "It's like an active foreign part in their bodies, which attracts attention, therefore it is prioritized in their temporal lobe," said Brugger. "It makes sense, in retrospect."

It is, however, deeply ironic that the foreign-feeling body part would be attracting more attention than the rest of the body. Contrast BIID with a condition called somatoparaphrenia, in which people often deny ownership of a leg or arm, or even an entire side of their body. The delusion often arises because the person has suffered paralysis of one side of the body and is also sometimes unaware of the paralysis. But in BIID, there is no such functional problem with the body part. So, the brain's increased attention to it makes sense only if you accept that the leg or the arm is not part of the bodily self constructed by the brain. BIID is also telling us that even though we can lose the sense of ownership over body parts, there is still an "I"—the self-as-subject—that experiences the lack of *mineness* of a limb. It's unfortunate that the estranged body part becomes an object of obsession, as would anything foreign that clung to one's body, an obsession that eventually leads some to amputation.

● ● ○

Visceral negative reactions are common when people first hear about voluntary amputations. About fifteen years ago, when media attention to BIID happened to be at a peak, bioethicist Arthur Caplan, then of the University of Pennsylvania, called it "absolute, utter lunacy to go along with a request to maim somebody."

More than a decade later, there is still a debate raging in the pages of academic journals about the ethics of voluntary amputations. Is it analogous to body-modifying cosmetic surgeries, such as breast reduction, as BIID sufferers themselves have argued? Some bioethicists say no, since amputation entails a permanent disability. Others point out that cosmetic surgery can also be disabling, as when breast reduction results in the inability to breast-feed. Some have compared BIID to anorexia nervosa as the best, if a somewhat imperfect, analogy, because both involve body-image discrepancies. According to this line of argument, amputations should be denied just as anorexics are sometimes fed against their will. The retort to this is that anorexics are clearly delusional about their bodies, as objective measures can show their body weight to be dangerously low. There is no accepted objective measure of a BIID patient's internal feeling of bodily mismatch.

The debate continues, partly because BIID is not a medically recognized disorder. There's also a lack of data about how voluntary amputations affect the lives of patients. Yet David's surgeon, an orthopedic specialist, has made up his mind.

Dr. Lee—which is not his real name—is in his mid-forties, friendly, with an easy laugh. He seems at peace with his secret practice. When a BIID patient first approached him six years ago, he'd had his doubts, so he researched BIID as thoroughly as he could and communicated with the patient for several months before deciding to do the amputation. He knew he would be risking his medical license. A religious man, he and his wife even prayed on it, eventually putting some of the onus of the decision on higher powers. "*God, if you think this is not right, then put some hindrance,*" he remembers thinking. "*I don't know what it is, but put some hindrance.*" So far, things have gone smoothly, and he's taking that as divine sanction.

Dr. Lee is convinced that what he does is ethical. He has no doubt that BIID patients are suffering deeply. On the question of whether to amputate to relieve their suffering, he invokes the World Health Organization's definition of health: a state of complete physical, mental, and social well-being and not merely the absence of disease or infirmity. As far as he can tell, people with BIID are not healthy; there is no nonsurgical cure in sight and no evidence that psychotherapy helps. Michael First, in his 2005 survey of fifty-two BIID patients, reported that 65 percent of them had seen psychotherapists, but it had no effect on their desire for amputation (though it's also true that half of them did not tell their psychotherapists about such desires).

Of course, there's also the question of whether BIID sufferers are psychotic or delusional. Again, the scientists who have studied these individuals say that they are neither. Dr. Lee insisted that his patients have not been psychotic (and as we'll see in the next chapter, psychosis in schizophrenia involves a profound alteration of one's experienced reality; no one with BIID I talked to said this was the case for them).

On the contrary, Dr. Lee said, many of his patients were high-functioning individuals, including a pilot, an architect, and a doctor. And for Dr. Lee the proof that BIID is a real condition can be found in the near-instant change he has observed in his patients after the surgery, which contrasts strongly with those who have to undergo involuntary amputations because of, say, a car accident. Involuntary amputations are traumatic to even the strongest of people, and those people can become severely depressed as a result. "Then you have these BIID people who crutch unbelievably after the first day after surgery."

Paul McGeoch, who has studied his fair share of BIID patients, has the same opinion. "They are universally happy. I have never heard of one who is not pleased to have a limb amputated," he said. But as convinced as Dr. Lee seemed, he repeatedly stressed to me: "I'll stop the moment I get my first patient who feels remorseful about the surgery. So far, none have."

If BIID were ever to be legitimized and voluntary amputations to become legal, Dr. Lee knows that his clandestine program would end. "I'd be so glad if ever that happened. I won't have to deal with the tension anymore," he said. "Right now, I'm torn between the tension of doing the surgery, and the tension of helping them." Then, in a momentary lapse of caution, he admitted that he would miss the surgeries: "Maybe that's the weirdo in me."

Would he miss the money, which amounts to about $20,000 per operation? The answer was an emphatic no. He said he made the same amount doing legal surgeries for foreign health tourists and that he had a flourishing local practice. He pointed out that his fee covered everything: hospital costs, payments to his fellow surgeons, even some meals and sightseeing for his patients. "You are not paying for the surgery. You are paying for all the risks involved," he said. "You have to keep everybody happy. We are not talking peanuts here. If this gets out, we all lose our licenses." He said it was a risk he's willing to take, as long as his patients are happy.

● ● ○

The morning that David was scheduled for surgery I went to meet him and Patrick in their hotel suite. We had flown thousands of miles to be there, in a crowded Asian city. Outside the hotel, the weather was hot and muggy, the traffic heavy. Luxury cars and jalopies jostled for street space alongside buses and two-wheelers. Diesel fumes stung my nostrils. A fetid stream wound its way between high-end hotels and office buildings. Inside the hotel, the wood-paneled suite was air-conditioned, hushed.

I had spent the night thinking about David's surgery, and all I had felt was anxiety. I imagined the fear that David must be experiencing: fear of surgery, fear of confrontations with family and friends, fear of disability. But that morning David himself showed no such emotions. He said he had moved beyond those worries. Instead, he fretted about the paperwork. Whom should he put down as emergency contacts? Should he divulge their addresses and phone numbers? Patrick suggested putting down the wrong numbers; maybe change a digit or two. "You'll have to get used to lying," he said.

Questions kept occurring to me. I asked David if he had been evaluated by a psychiatrist. Usually, Patrick recommended someone for surgery only after a psychiatrist confirmed that he or she was suffering from BIID. David said no. Patrick had used his own judgment in recommending him to the surgeon, saying that he saw himself in David—the same agony, the same mental torture. Plus, David couldn't afford a psychiatric evaluation. He had to scrape and scrounge and go deeply into debt to come up with the $25,000 needed to cover the surgeon's fees, the airfare, and ten days of hotel accommodation for two.

Dr. Lee had agreed to the surgery based on Patrick's recommendation. The two had been working together ever since they met via the

BIID network about six years ago. David was thankful for Dr. Lee's help. "As you know," he told me in the hotel room, "I was in a DIY mode, where I was going to hurt myself." Suddenly, David started sobbing. Patrick consoled him; David apologized. "Every time I talk about hurting myself, it makes me cry," he said. David again expressed certainty that if the surgery didn't go through, he would attempt cutting his leg off himself. "I can't go on any longer."

The surgeon picked us up in the early afternoon. Given that David's procedure would require subterfuge to get past hospital staff and nurses, Dr. Lee appeared surprisingly calm. "Have to be," he replied when I asked him later about his demeanor. "Cannot show the patient that I'm nervous." He drove us to his house, ushered us into the living room, and asked David to sit down.

Dr. Lee laid out the plan: he would get David admitted to a hospital, saying he needed surgery for a vascular disorder. The unwitting staff would then prepare the patient for an ordinary operation—and then, under the surgical lights, Dr. Lee would say that the leg needed removing and conduct the amputation. Inside, the anesthesiologist and other surgeons would be in on the plan; the nurses would not know.

In his living room, Dr. Lee laid an old garment on the floor, and set David's foot on it. Working swiftly, he bandaged the foot, ankle, and calf as a precaution. It wouldn't do to have curious hospital staff see that the foot was healthy. He wrote the admitting order on his prescription pad and instructed David in the sequence of symptoms he ostensibly had endured over the past few days: pain, followed by some cramping, and eventually numbness. This was for the benefit of the hospital admissions staff. The diagnosis that these symptoms implied would give Dr. Lee the option of amputating during surgery, a judgment that could not be questioned by anyone who wasn't in the operating room.

We drove to a small hospital on the outskirts of the city. The high-rise hotels gave way to low-slung buildings and occasional homes with makeshift tin roofs alongside unpaved muddy alleys. The hospital itself was on a major road lined with an odd assortment of shops: a butcher, a pawnshop, an electronics repair outfit, and a hairdresser who promised safe and effective hair straightening.

Dr. Lee was not on staff at this hospital; like many doctors in private practice he had surgical privileges at a number of different hospitals. He dropped us off outside. David, now on crutches, would have to get past the hospital staff. Would they buy his story? We walked into the emergency room. It was a simple affair. Ten iron-frame beds and mattresses covered with spotless sheets were separated by thick curtains. This was not a high-tech, First-World ER, but it was clean and functional.

A nurse asked David to sit down and asked what was wrong with him. He gave her Dr. Lee's admitting order. The attending physician, a bespectacled man in a blue-striped shirt with a stethoscope around his neck, took the order and frowned as he read it. He leaned over the counter to take a look at David's leg. He noted the bandaging and asked if David had suffered an accident. No, said David, and he quietly repeated the sequence of symptoms. The man got up and walked away.

David was subdued. Patrick, wearing his prosthetic leg, appeared to be feeling fine; he had been through this scenario many times. David, beneath his quiet demeanor, was nervous, as was I, even though I was just an observer. My mind raced through all that could go wrong. What if the attending doctor asked more questions? What were the three of us, two of whom were on crutches, doing in this part of the world? What if they called the police? Then, once David was done filling in the paperwork, a nurse brought him a wheelchair. She inserted a catheter into David's left hand and hooked it up to an IV bag hanging off of a pole on the wheelchair. She left. I looked at Patrick. "I can't believe it's really happening," he whispered in relief. A male nurse came in, and we got up and followed him as he wheeled David up to his hospital room. They had bought the story.

In the hospital room, we sent the surgeon a text message to say that David had checked in. Dr. Lee told me later that his own nervousness usually sets in the moment he receives that message. Now all was in motion.

As we waited in the hospital room, Patrick started giving David advice about life as a leg amputee. Don't ever close your eyes when you're standing without support, he said. You'll lose balance and topple over. Always carry powerful painkillers: stumbling and landing on your stump can be excruciating.

A nurse came in and informed David that the doctor would operate in a few hours, then left us alone again. We counted the saline drops dripping into David's veins: about twelve drops per minute. I asked David about his cover story for when he got back home. He said he would tell people at home the story he had told the hospital. Dr. Lee would provide him a full medical report to take back. Patrick recalled his own cover story: he'd picked up a rapidly spreading infection called St. Anthony's Fire while on vacation; the rampant infection turned the leg gangrenous, leading to the amputation. It had worked well for him. Then Patrick told David to do something one last time, for once the operation was complete, he would never be able to do it again: cross his legs. David did so. It was as if we were mourning an impending loss with a collective moment of silence.

Soon, two male nurses wheeled in a gurney. David lay down on it and he was taken away to surgery. Patrick gave him a thumbs-up. I didn't know what to say, so I just muttered, "Good luck," under my breath.

● ● ○

The hospital has gone quiet, and empty benches line the dimly lit corridors. Only the operating room shows signs of activity. David lies on the operating table, anesthetized and oblivious to pain. An overhead surgical lamp illuminates his upper thigh. Dr. Lee picks up a scalpel and makes a long, deep incision precisely where David had requested, in a leg that is athletic, muscular, healthy. The surgeon swiftly cuts through muscle, working hard. He cauterizes the smaller blood vessels as he goes while keeping clear of the large veins, arteries, and nerves. He pulls at the nerves, teases them free of the surrounding muscles, cuts, then lets go. The nerves retract into the soft tissue of the upper thigh like rubber bands. He clamps the large blood vessels, snips them, and ties up the proximal and distal ends, the proximal end three times for peace of mind. The surgery is taking longer than anticipated because the leg is so robust, engorged with blood. Finally, he slips a wire saw under the femur. An assistant presses down on the leg. Dr. Lee begins sawing and soon pulls the saw through the strongest bone in the body. He then attends to the blood vessels, nerves, muscle, and skin on the underside of the bone until the leg is finally detached. It is time to suture. First he sews up the muscles, then the fascia, the strong fibrous tissue surrounding the muscles. Suturing the fascia correctly is critical, because mistakes can lead to muscle herniation, a serious complication. Finally, the surgeon stiches up the skin and the subcutaneous tissues. Where once there was a leg, only a stump remains.

● ● ○

I wasn't in the operating room that night. But I did walk the empty corridors outside, trying to peer discreetly over the frosted glass panes of the doors leading into the OR. I have thought about the surgery (which Dr. Lee had described to me in detail) many times since. Each time I have felt fear and sadness. Here was a perfectly healthy man with a perfectly healthy leg, yet he went under the knife voluntarily, in a foreign country. He trusted a surgical team that worked under a cloak of deception. How much must a man suffer to come to this: lying by himself on an operating table, attended only by strangers, in a small, obscure hospital thousands of miles from his home in America?

● ● ○

Patrick was asleep when I heard the knock on the door. It was more than three hours since David had been wheeled away. It was a male nurse in surgical robes and rubber gloves. He turned to Patrick and said, "The leg has to be buried as soon as possible." He needed money to take care of the burial. Patrick handed him some cash. "Do you want to see the leg?" the nurse asked. "It's already in the box." Patrick didn't. The nurse left. "Well, he's an amp now. I'm glad," said Patrick. "It's what he wanted. It's what he needed."

Dr. Lee appeared soon after. The surgery went well, he said, though it had taken longer than usual. David was fine and lay asleep in recovery. Dr. Lee offered to give me a ride back to my hotel, and I accepted. During the ride he talked about David's long operation. "His muscles were well built," said Dr. Lee. "They contract and they also bleed more. You have to be careful." Still, there was the satisfaction of a job well done. "What's fascinating is that you can really see the transformation," he said. He meant the change in the demeanor of his BIID patients after surgery. "You'll notice it tomorrow."

The next day, I couldn't wait to get back to the hospital. I bought a bar of bittersweet chocolate for David and hailed a taxi. When we arrived, I walked in through the front door, past the ER, and paused for a moment at the frosted glass doors of the operating room. Then I went to David's room and knocked on the door. Most patients would be flat on their backs recovering after such a major surgery, but David was sitting up in his bed, his stump heavily bandaged and covered in white gauze. He was still on an IV. Tramadol, a narcotic-like analgesic, was dripping into his veins. He was tethered to a urine bag. He looked tired, but then it was only twelve hours since the operation. I shook his hand and gave him the chocolate. David opened the wrapper, broke off a piece, and began to eat. He sat on the hospital bed as if nothing dramatic had happened the night before. Our conversation eventually wore him out. He fell asleep.

When I returned the next day, the IV and the urine bag were gone. A pair of crutches lay next to David's bed; he had already crutched to the bathroom and back, just as the surgeon had said he would. He smiled and laughed easily as we spoke. The tension that had lined his face all the time I had known him was gone. I sensed relief, happiness.

Months later, I exchanged emails with David. He said he had no regrets. It was as if for the first time in his life, David was whole.

4

TELL ME I'M HERE

WHEN YOUR ACTIONS DON'T FEEL LIKE YOUR OWN AND WHAT IT DOES TO THE SELF

What gives me the right to speak of an "I," and even of an "I" as cause, and finally of an "I" as cause of thought? . . . A thought comes when "it" wants, not when "I" want.

—Friedrich Nietzsche

For any true grasp of delusion, it is most important to free ourselves from this prejudice that there has to be some poverty of intelligence at the root of it.

—Karl Jaspers

March 10, 2013. It was a bitterly cold day in Bristol, England, much colder than it was in London, which is a two-hour train ride due east and from where I had just arrived. I met Laurie and her husband, Peter, at the Bristol train station. We were to go and see the parking garage where Laurie had tried to jump and end her life on a similarly cold day in November 2008.

Peter drove us to the garage and up the ramp that spiraled steeply to the terrace of the eight-story building. "You are not going anywhere near the edge," Peter said to Laurie. "You are not going to tempt fate." Laurie seemed far less concerned. She exclaimed, "Wheeeeee!" like a kid on a roller coaster as Peter climbed the ramp at a fair clip.

We parked on the seventh floor and climbed up to the terrace. The wind stung. For a few minutes, Laurie struggled to find the spot she had intended to leap off of. Nothing looked familiar. Even the parapet was too high. "This is impossible for me to climb," she said. "I think they changed it, to make it less easy to climb." But the concrete parapet looked uniformly old, nothing seemed added on. We kept searching.

We found the place. It was near the very top of the same spiral ramp we had driven up. The ramp had inner and outer parapet walls. That fateful November day, Laurie had first peered into the inside of the ramp. The ground below had been muddy (it was filled with gravel today) and she had decided it would be too soft to kill her. She then walked over to the chest-high, foot-wide outer parapet and somehow climbed onto it. Had she jumped, she would have landed on concrete.

Today, when you stand at the wall, you see a modernistic fifty-foot-high sculpture in front, a column clad in slate, with an umbrella-like disk of solar panels near the top, above which are twin, twisting vertical blades of a wind turbine. "I remember looking at that," Laurie told me. "They were building that in 2008."

The sculpture stands in the middle of a long traffic island. On the far side are multistory brick buildings, beyond which you can see the tiered tower of St. Paul's Church, known in Bristol as the Wedding Cake Church. Even with suicide on her mind, Laurie had stood admiring the view. It had also given her pause to ponder the jump. Would it kill her or just paralyze her? As she contemplated the outcome, a man saw her from below and called out: "Are you all right?" Laurie did not answer. "I guess he called the police," she told me. The police came up and rescued her. They took her to the nearby police station, where she was sectioned under the UK's Mental Health Act—in a holding cell for twenty-four hours.

To this day, Laurie thinks it wasn't her decision to attempt suicide. "I was under the influence . . . of some force," she said. "I wasn't the one making that decision. Someone was trying to push me off the edge."

Soon after that incident, Laurie was diagnosed as suffering from schizophrenia. But the knowledge hasn't changed her sense of how she felt the day she tried to jump. Sitting in a Starbucks inside the shopping center next to the parking garage, she continued to voice skepticism that the thoughts that told her to jump were her own. "I still wonder if it is outside of me," she said.

● ● ○

About a month later, I was attending a conference on "Hearing Voices" at Stanford University. The first speaker had finished her talk on musical hallucinations, and was taking questions. An audience member read out a question that someone named Sophie had posted on Twitter (the talk was being streamed over the Internet). Suddenly, a woman sitting near the front put up her hand. As the puzzled speaker looked at her, the woman said, "Sorry, I'm Sophie." The audience dissolved into laughter.

I had a more complicated reaction than that of the audience. I had come to the conference to meet Sophie (who is from Chicago), so the fact that she was posting on Twitter had dismayed me. Was she watching the talks remotely? Hadn't she turned up at Stanford? Seeing her sitting in the room was a big relief.

I first learned about Sophie from Louis Sass, a professor of clinical psychology and a schizophrenia expert at Rutgers University in New Jersey. "She's the most articulate person with schizophrenia I have ever met," Sass told me. Years ago, before her own tryst with schizophrenia, Sophie contacted Sass because she had found his work interesting. Sass has been arguing for decades that schizophrenia should be viewed as a complex disturbance of the self and self-consciousness, and the view resonated with Sophie, whose mother had suffered from schizophrenia. Then, one day, Sass received an email from Sophie, which he recollects as saying, "Gee, funny thing happened . . ." Sophie, it turned out, had had a psychotic breakdown herself.

Sophie grew up with a mother who suffered from psychosis (a condition in which one's sense of reality is profoundly altered). With the hindsight of maturity and training in psychology and philosophy, Sophie can see her mother's paranoia and erotomania ("she was convinced people were in love with her") for what it was: the outcome of severe schizophrenia. But as a four-year-old, Sophie knew no better. Her mother would drive Sophie and her brother to grocery stores but would refuse to go in herself. Instead, she would send the children to get the groceries and even pay for them. "When you are a four- or five-year-old child, getting a whole cartful of groceries and paying for them with a check that your parent has pre-signed was very strange," Sophie told me. "But at the same time, I thought, Oh, that's just how she is."

Her mother's paranoia manifested in other ways. For instance, when strangers, or even the postal carrier, came to their house, the family would shut all the windows and hide. "I thought it was very normal," said Sophie.

It was around the time that Sophie entered junior high that she realized her mother, and their family life, wasn't normal. Her mother's paranoia had been exacerbated. She thought a recording device had been implanted in her uterus, and even in their dog, and that the whole house was wired. She would ask her kids to walk down a block, away from the house, before she would talk to them.

If that wasn't difficult enough, Sophie's family history of schizophrenia went further. Her mother's first husband had had a schizophrenic breakdown when studying philosophy, and was institutionalized at a state hospital in California. "We grew up in fear of him," recollected Sophie. "[My mother] thought that he was going to get out of the hospital that he had been committed to, and come and find us, and that he wanted to kill her. I have no idea if that was in any way grounded in reality. So we grew up in fear of him, but at the same she very strongly romanticized his brilliance, his genius. Our house was full of his philosophy books."

Kant, Hegel, Heidegger, and Karl Jaspers filled the shelves. Sophie even got to read this man's diaries, which documented his descent into madness.

Through all of this, Sophie negotiated her childhood just fine, developing an intellectual and academic bent of mind. She turned down a Cornell scholarship and went to Nepal to work with an NGO, and then spent a year and a half in Japan. She returned to the United States and went to the University of Oregon in Eugene to study continental philosophy. One of her advisers was John Lysaker, who has written extensively on schizophrenia, psychosis, and the self. During her senior year, still blissfully absent of any symptoms of psychosis herself, Sophie wrote to Louis Sass. She was intrigued by his ideas on schizophrenia, the attendant "madness," and the parallels he saw in modernism.

● ● ○

"If you want to find a good analogy for many schizophrenic experiences and symptoms, an excellent place to look is in the avant-garde modernist and postmodernist art of the twentieth century," Louis Sass told me. "That is not to say anything as silly as modernism is schizophrenic or that schizophrenia is modernist necessarily, but there's a structural parallel which offers quite a different way of understanding, often in great detail, a lot of what is really going on in schizophrenia."

An unusual confluence of life events led Sass to this view of schizophrenia and to his 1992 book *Madness and Modernism.* One was his training in modernist literature. As an English major at Harvard in the late '60s, he was drawn to modernism, wrote his thesis on Nabokov ("who was kind of a modernist"), and keenly studied the poetry of T. S. Eliot and Wallace Stevens. Schizophrenia was also a hot topic then. Scottish psychiatrist R. D. Laing had written a provocative book on the subject, *The Divided Self.* Sass took a course at Harvard for which Laing's book was required reading. And around that time a close friend of his developed schizophrenia.

Almost four decades later, sitting at the kitchen table in his Brooklyn brownstone apartment, Sass recounted his friend's descent into the cauldron of schizophrenia. There were signs even in high school that his friend was unusual. Those who develop schizophrenia typically go from being premorbid (before there are any clear indications of impending psychosis) to prodromal (at the cusp of psychosis) to full-blown psychosis. "His premorbid personality, to use the technical term—I certainly didn't think of him that way, he was my friend—was in retrospect typical of a certain kind of person with schizophrenia," said Sass.

His friend was unconventional and fiercely autonomous (an attribute that would prime Sass to question the standard view that mental disorder always involves lessened autonomy). "We 'normals' were so incredibly conventional from his point of view," said Sass. "So cowardly, in a way. . . . You wouldn't dare stand on your head here in my house, for example. He would have, if he felt like it, as a manner of speaking. He would do things that were outrageous. He wasn't afraid of anything."

Once, in a school cafeteria, his friend picked up the fish from his plate and lobbed it high into the air and clear across the hall toward the teachers' table. All this could be described as behavior "motivated by a certain kind of oppositionality, contrarianism, insistence on autonomy, contempt for the normal," said Sass. Not entirely unusual for adolescent boys of his age, perhaps. But "there was something different about my friend's way of manifesting it . . . so extreme that one has to call it, whatever the word means, 'insane' in a way."

His friend eventually became psychotic. "My sense of what it was from knowing him and knowing him very well, from before he became psychotic and after, didn't fit with the common images [of schizophrenia]," said Sass.

Schizophrenia was originally called dementia praecox, a term coined in the 1890s by the German psychiatrist Emil Kraepelin. It was Swiss psychiatrist Eugen Bleuler who renamed it schizophrenia in 1908. Dementia praecox, or premature dementia, posits, among other things, intellectual disability. Another now out-of-favor psychoanalytic view of schizophrenia was one in which the person regressed to an infantile state, robbed of the maturity of an adult. Yet another stereotype, popularized by the antipsychiatry movement and some of the literary avant-garde, was of the schizophrenic as a romanticized wild man, in touch with his deepest desires and instincts.

Sass and his friend went to different colleges. Sass went to Harvard, and would go on to do his PhD in psychology at the University of California, Berkeley, and his internship in clinical psychology at the Cornell University Medical Center–New York Hospital. Meanwhile, his friend's schizophrenia worsened. He dropped out of college, and eventually committed suicide. The experience marked Sass.

Back in his apartment, Sass cast his mind back to when he had gone to see his friend after he had had psychotic breakdowns. On one occasion, Sass found him obsessed with dancing on one foot—something he had been working on for many weeks—now demonstrating his talent inside his mother's garage. But there seemed to be no further purpose to his endeavor, no desire to impress anyone, no desire for personal gain or any usual sort of narcissistic satisfaction.

"He was an extreme, and from any normal point of view insane, devotee of autonomy. I'm not trying to say it's a better way to live, obviously, but it offends me deeply at some ethical level, and at some aesthetic, intellectual level as well, that these things would not be recognized for what they are," said Sass. "Scientifically, it's a failure to recognize the true nature of the phenomenon, in all its sometimes paradoxical complexity."

What Sass, then, is arguing for—and he's not the only one—is for psychiatry to move away from describing schizophrenia so exclusively in terms of deficits—lacking this, lacking that—and to think of it positively. By "positive," he does not mean good. He means to recognize what it feels like to be schizophrenic, to understand its phenomenology, not just to note the failure to conform to cultural standards.

One way to understand schizophrenia, Sass argues, is to look toward modernism in art (the cubism of Picasso, the dadaism of Marcel Duchamp, and the surrealism of Giorgio de Chirico and Yves Tanguy, for example) and literature (Franz Kafka and Robert Musil, T. S. Eliot and James Joyce, to name a few). Such art can give us a sense of what the schizophrenic experience might entail. In the various traits of modernism, as well as postmodernism, Sass sees threads of what he termed "hyperreflexivity" (a kind of exaggerated self-consciousness that takes what would normally be the implicit medium of our experience and turns it into an explicit target of excessive focus and attention) and also of alienation. "Instead of a spontaneous and naïve involvement—an unquestioning acceptance of the external world . . . and other human beings, and one's own feelings, both modernism and postmodernism are imbued with hesitation and detachment, a division or doubling in which the ego disengages from normal forms of involvement with nature and society, often taking itself, or its own experiences, as its own object," he wrote.

● ● ○

Laurie can recall the feeling of her first major encounter with schizophrenia. It was Bonfire Night, during the fall of 2005. Across the country, fireworks were being lit to celebrate events of November 5, 1605, when the police thwarted a plot to blow up the parliament building in London. Laurie was seventeen, in boarding school in Canterbury, England. She watched the fireworks display and then came back to her room and sat down in her chair. She felt strange. As if something were controlling her, possessing her, an outside force. She sat for a couple of hours, doing nothing, just preoccupied with the strangeness. Then she picked up an art knife and cut her left hand. And she went to sleep. She woke the next morning and cut herself again, this time a lot deeper. The bleeding wouldn't stop. "Somehow I just snapped back to reality, and realized, Oh, gosh, I have cut myself," she told me. She and a friend rushed to seek medical help.

That incident was the first serious realization that something was amiss. The Bonfire Night incident brought into focus something she had begun to feel a few months earlier: paranoia that she would be deported from the UK back to her home country, though there was no cause for concern in real life. Until then, she had dismissed those fears. But the thoughts became more frequent and insistent. They had an almost acoustic quality, an "external physical quality," as they crowded her head. They were linked to deportation. Their refrain was unfailingly familiar. " 'Nobody will miss you if you go, you are useless, you are a failure,' that kind of thing," she told me.

By March 2008, Laurie had cut herself more than a dozen times. That was when she and Peter, then her boyfriend, took a trip abroad to meet her parents. One night, when everyone else had gone upstairs to bed, Laurie showed Peter the scars on her hand.

" 'Oh, dear,' I think were my precise words," Peter told me while we all sat down for dinner at the Hole in the Wall pub in Bristol.

"You said, 'Oh, Jesus,' " Laurie corrected him.

"Fair enough," said Peter.

Soon after her talk with Peter, Laurie began hearing voices. She remembered the month: May 2008. It was unclear whether it was the voice of one person or three, for the voices seemed to echo inside her head. But it was a middle-aged voice speaking in a British accent. The woman or women spoke directly to her, telling her to cut deeper, to kill herself. These voices, speaking in the second person, as it happened, delayed Laurie's diagnosis of schizophrenia, for which she blamed Kurt Schneider, an early-to-mid-twentieth-century German psychiatrist. Schneider had cataloged a set of what he called first-rank symptoms for diagnosing schizophrenia. Among these are third-person auditory hallucinations, in which the voices talk to one another about the patient. Though Laurie displayed some other first-rank symptoms (thought insertion, or the feeling of alien thoughts in her head, and primary delusion, a delusion that appears unbidden and without precursors, which in Laurie's case was the feeling that her surroundings had an inexplicable strangeness), her psychiatrist, idiosyncratically and mistakenly sticking to Schneider's old ideas, regarded the presence of second-person voices as uncharacteristic of schizophrenia, and more an indication of psychotic depression (even though we now know that many people with schizophrenia do hear voices speaking to them directly).

The staggering array of symptoms in schizophrenia complicates diagnosis. The symptoms are usually classified as positive (delusions, hallucinations), negative (apathy, emotional flatness), and disorganized (such as jumbled-up speech). Diagnosis often involves ruling out other disorders before settling upon schizophrenia. In Laurie's case, it meant being diagnosed first as suffering from depression, then from borderline personality disorder. Meanwhile, her attempts at suicide got more serious. She once overdosed on eighty tablets of acetaminophen, and suffered two weeks of vomiting. Soon afterward she tried to jump off the eight-story parking garage. And around that time, a psychiatrist diagnosed her with schizophrenia.

Sometime in early 2009, her condition worsened. She tried to kill herself again, this time with an overdose of her antipsychotic medication. Even her very sense of being a person was threatened. "During that period of intense symptoms, I thought my whole self disintegrated and dissolved; I didn't have one," she said. For instance, if she held out her hand, she would feel it going farther and farther away. "My sense of self, bodily self or psychological self, or a combination of the two, was just permeating outwards," she said. "Even when I was just sitting, I'd think I was just transparent, almost. Not physically, obviously, metaphorically."

● ● ○

Sass and his colleague Josef Parnas, a psychiatrist at the University of Copenhagen, Denmark, think that the answer to the conundrum that is schizophrenia lies in the self. Scientists have long struggled to come up with a unifying hypothesis for schizophrenia. What possible common mechanism could underlie the diversity of positive, negative, and disorganized symptoms? Could it be a disturbance of the very underpinning of our being, a disturbance of our sense of self?

To explain schizophrenia, German psychiatrist Karl Jaspers coined the term *Ich Störungen*, which literally translates to "ego disturbances." Jaspers used the term to signify how the core symptoms of schizophrenia all have something to do with a disturbance of the boundary between the self and the other, the self and the outside world.

Sass and Parnas think that schizophrenia is the result of an even more basic disturbance of the self. The duo's thinking owes much to a long tradition of mostly European phenomenologists—phenomenology being "the study of 'lived experience.'" These phenomenologists include, notably, Edmund Husserl, Martin Heidegger, Maurice Merleau-Ponty, and Jean-Paul Sartre. It's through the analysis of the lived experiences of patients that Sass and Parnas arrived at their thesis: schizophrenia involves the disruption of a basic form of selfhood. To understand their point of view, we need to treat the self as a layered entity. There is the by-now-familiar narrative self—the stories we tell ourselves (and others) about ourselves, an identity that spans time, from the past to the future. But even before the emergence of the temporal storyteller within us, there's the self-as-subject that is able to reflect upon aspects of itself, these aspects constituting the self-as-object (our narrative would be one such aspect, or object, for the self-as-subject). Sass and Parnas are targeting the self-as-subject: it's "the fact that I feel that I exist now in this moment, that I feel a sense of being a [subject], a sense of being the thing to which things are happening, and from which acts emanate," said Sass. They call this *ipseity* (*ipse* is Latin for "self" or "itself").

During our meeting, Sass displayed extemporaneous literary eloquence as he described the concept further. "Ipseity is that from which the fiats of the will emanate, and toward which perceptions come. It is the implicit sense of feeling that you are here. But of course, you don't think about that directly. It's a feeling, and it's of its essence that it *not* be the object of awareness," he said. "You might say that it's the nowhere from which will emanates, the nowhere to which perceptions arrive; that's more or less how William James described it."

"That it *not* be the object of awareness . . ." It's this assertion that holds the clue to Sass and Parnas's idea of what happens during schizophrenia. The disorder, they argue, involves a kind of hyperreflexivity, an undue amount of attention paid to aspects of oneself that otherwise just exist without being the focus of attention. "It's a subtle but crucial phenomenological difference between moving your arm and taking the movement of your arm as the object of your attention," said Sass. "Those are very different things."

Sass and Parnas posit another seemingly contradictory disturbance of ipseity that they think is present in schizophrenia. It's what they call "diminished self-affection": a reduced sense of being an entity to which things happen, of being an entity that is the subject of awareness. Sass writes, "This experience of one's *own* presence as a conscious, embodied subject is so fundamental that any description risks sounding empty or tautological; yet its absence can be acutely felt."

Laurie could attest to that. In the days leading up to her suicide attempt from the top floor of the garage, she felt an intense emptiness. "When I was in that state, I just thought there was so much nothingness around me, inside of me, I couldn't function," she told me. "I thought if I couldn't function, what's my worth? I might as well be dead."

Sass and Parnas argue that when ipseity is disturbed, the basis of our very being is eroded, making it fertile ground for psychosis and releasing all sorts of strange experiential possibilities.

● ● ○

During the early phase of her psychotic break, Sophie remembered noticing subtle changes too. Sophie told a friend, who was French, about how she was seeing the world as particles, and how it felt as if the mere act of blowing on a building would disperse it into thin air. "To this day, I don't know where the mistranslation occurred, whether it was on her end, or her expressing that in her French-English to a professor, but somehow they decided that I was planning to blow up a building," Sophie told me. She was banned from the philosophy department where she was a student, and threatened with arrest if she showed up on campus. Sophie went to the campus anyway to see her adviser, who refused to meet her and slammed the door in Sophie's face. Sophie was initially temporarily suspended, but a year and a half later she was permanently expelled from the department.

Even before that happened, Sophie was struggling as a student. She would find herself unable to talk, sometimes for hours on end, despite being perfectly capable of formulating thoughts and sentences in her mind. The words just wouldn't come out. That was incredibly inconvenient, given she had office hours as a teaching assistant, or had to attend classes as a doctoral student. Unable to afford good psychiatric care, Sophie went to a psychiatric hospital in Chicago meant to help poor and low-income patients. The experience was scarring. The intake nurse told the friend who had accompanied Sophie, "I'm not the one doing the official assessment, but from what you have told me it seems as plain as day that she's a schizo." The comment stung. "I was right there," Sophie told me, indignation rife in her voice even years later.

The hospital locked her up, in a spartan room, surrounded by others who were suffering from various mental-health problems, including substance abuse. Sitting among patients who were walking around screaming and yelling, Sophie was unnerved at where she found herself. "It was disturbing to me, from that perspective, although I had grown up with my mother and was used to dealing with her," she said. Her friend, horrified at the way Sophie was being treated, helped her escape from the lockup.

Fortuitously, Sophie discovered a well-funded program that focused on first episodes of psychosis. She called the clinical director. The response was immediate. "She said, 'I want to see you at seven a.m. in the morning,'" recalled Sophie. "She was incredibly reassuring and nice, and it was just night and day." Sophie enrolled at the program for intensive treatment. But despite talking with the clinical director a number of times during the week and taking antipsychotic medication, Sophie wasn't convinced of her psychosis, partly because she thought her altered view of the world made sense, thanks ironically to her training in philosophy. While her mother's madness had been "profoundly irrational, conspiracies, plots, and things going on," Sophie's own perception of the world as insubstantial, where solid boundaries melted away into an amorphous whole, did not seem unrealistic. Solid objects were illusions. Even the reality of people existing as individuals was tenuous. "That felt entirely in line with the types of questions that philosophers have been asking for centuries," said Sophie.

Meanwhile, her schizophrenia was having profound effects on her being. Her sense of a barrier between her internal and external worlds had dissolved. "Suddenly, it was as if my entire interior life was exposed to everyone," she said. During her sessions with a psychiatrist, she was constantly being asked if she was getting messages from, say, the radio, or whether she was hearing voices. While she wasn't getting messages or hearing voices, Sophie felt compelled to know whether she was psychotic or not. She began fixating on objects to see if they were communicating with her, and started focusing on her own thoughts. "This is what Louis [Sass] would call hyperreflexivity in the most self-conscious sense—the more I concentrated on my thoughts, the more objectified they became, the more I started to hear auditory elements to things," said Sophie.

Schizophrenia has also changed Sophie's relationship to her own body. "My hands never look like my own hands," she said. "There must be some sort of split-second gap between the movement of my hand and me registering that as my own action, or a self-initiated action."

What Sophie experienced and continues to experience is a disruption of what is called our sense of agency. It's that part of our sense of self that makes us feel that we are the owners of our actions. If I lift a glass of water, I know that I'm doing the lifting. Can something we take so much for granted go awry? And could it cause psychosis—the perception of a distorted, nonexistent reality? The answers have their roots in experiments with fish, flies, and eyeballs that began in the early nineteenth century.

●●○

Move your eyes left to right, back and forth. What happens to the scene that you are looking at? If all's well with your visual system, then you should see what's to your left or right, but the scene you are looking at should hold steady despite the fact that your eyeballs are moving. But think about this for a moment. As far as the brain is concerned, the signals falling on your retinas could be due to either the motion of your eyeballs or something moving in your visual field. How does it know which?

Charles Bell and Johannes Purkinje, back in the 1820s, independently figured out that the answer to this question was telling us something very important. When you move your eyes normally, the brain cancels out the expected movement of the image—because it knows it initiated the eye movements, thus keeping the image steady. But when something is moving in the visual field, there is no such cancellation, and we perceive motion.

Then, in 1950, Erich von Holst and Horst Mittelstaedt carried out an experiment that illustrated this rather more bizarrely. They twisted the neck of the blowfly *Eristalis*, turning its head upside down: "Eristalis has a slender and flexible neck which can be rotated through 180° about its longitudinal axis. If this is done, and the head glued to the thorax, the positions of the two eyes are reversed," they wrote. The fly demonstrated truly strange behavior: in darkness, it acted as if nothing was wrong and moved normally, but under lights, it started going around and around, either clockwise or counterclockwise, choosing the direction at random once the lights came on and sticking with it. The same year, and independently, neurobiologist Roger Sperry did something similar. He surgically rotated the left eye by 180° in southern swellfish (*Sphoeroides spengleri*), and blinded the right eye ("Its small size, loose, scaleless integument, and general hardiness make this fish suitable for experiments involving surgery," wrote Sperry). Once the fish recovered from surgery, it too would circle either to the left or the right.

Von Holst and Mittelstaedt came up with the term *Efferenzkopie*, or *efference copy*, to explain what was going on. Sperry used the term *corollary discharge*. The essence of the idea was the same in both cases. The animal's brain generated a command to move. A duplicate of this signal was sent to the visual center. The nervous system would use the copy to compare the expected movement with the signal of the actual movement and use this comparison to stabilize the animal's motion—a kind of feedback mechanism to ensure that it was moving accurately in the intended direction. But if the head or eyes were twisted around, the feedback reinforced errors instead of correcting them, causing the animal to move in circles.

What could this possibly have to do with schizophrenia, psychosis, and the self?

In 1978, Irwin Feinberg of the VA Hospital in San Francisco tackled this question head on. Experiments until then had shown that motor actions could produce a corollary signal or copy, at least in simple animal models. Could such signals be used to distinguish self from non-self? Say your arm moved. Could the brain use the corollary signal to tell whether the arm moved because *you* tried to move it, or whether it moved due to an external cause?

The question is not as weird as it sounds. Before Feinberg published his paper, the Canadian neurosurgeon Wilder Penfield had written about experiments in which he would stimulate the motor cortex of patients who were undergoing exploratory surgery for treatment of epilepsy. The stimulation would cause the arm to move. But the patient insisted that he had not moved the arm, rather that Penfield had caused the arm to move. Because the patient had not willed the motion, no motor commands were willfully initiated and there would have been no corollary signal; so, the hypothesis goes, the brain attributed the movement not to the self but to an external agency. Feinberg eloquently argued: "The subjective experience of these discharges [or signals] should correspond to nothing less than the experience of will or intention."

And Feinberg went further. What if corollary signals were not limited to motor actions but also to thoughts? Could this be the mechanism for making a thought seem as if it belonged to oneself, rather than to someone else? Feinberg suggested this might be the case. He even attributed auditory hallucinations to malfunctions of this "corollary discharge" mechanism. Indeed, he posited that such malfunctions lay behind some of the strange symptoms of schizophrenia, even the blurring of boundaries between self and non-self, the kind experienced by Laurie and Sophie and countless other sufferers of schizophrenia. "Thus, if corollary discharge, in permitting the distinction of self-generated from environmental movement, thereby contributes to the distinction of self and other, its impairment might produce the extraordinary distortions of body boundaries reported by schizophrenic patients," wrote Feinberg.

● ● ○

During the depths of her psychosis, Laurie would hear voices a few times a week, women telling her that she was useless, a failure. Her husband, Peter, could tell when she was hearing voices. "She would look vacant and gaze off into space. Or she'd respond to the voices;

she'd say something completely out of the blue," said Peter. "You would instantly [know] she was responding to the voices."

Peter would actually engage with the voices through Laurie. She'd tell him that the voices were saying she was a failure. "Why do they think that?" Peter would ask. The voices would respond, "Because you failed to get your degree." And Peter and the voices would argue back and forth, with Peter pointing out to the voices that Laurie hadn't failed her degree, she had merely taken a year off from university (which she had, to cope with her illness). These episodes would last for half an hour, sometimes an hour, and eventually the voices would subside.

Laurie comes across as deeply introspective and analytical. These are traits that forced her to question her condition. She wanted answers. Was she crazy? Her inward journey resulted in two papers that she wrote when she was still a student struggling with schizophrenia. In one of the papers, she ends with a plea to psychiatrists to pay heed to what the patient is saying. Her experience with the psychiatrist whom she saw immediately after her attempted jump off the eight-story garage is illuminating. She explained to him that she had watched herself from a detached, third-person perspective as she tried to commit suicide. She had not been herself. Her psychiatrist dismissed her observations by saying, "You certainly communicate your distress clearly." It's in response to such indifference that Laurie implored psychiatrists to recognize the "unwanted new reality" that schizophrenia foists on people, which could help "rescue the sufferer from his [or her] isolation."

● ● ○

A mainstream psychiatric idea for why schizophrenics have to confront this painful reality relies on the notions of self-checking corollary discharges. The idea that an animal might distinguish aspects of self from non-self via this mechanism has been tested even at the level of single neurons. Singing crickets (*Gryllus bimaculatus*) chirp at an astounding 100 dB SPL. Their chirps are synchronized with their wing movements, with the crickets generating pulses of sound as they close their wings. Amid this cacophony of sound, how does a cricket—whose ears remain sensitive at all times—distinguish between its own chirps and external sounds? It turns out that there's a single interneuron that manages this task. This corollary discharge interneuron (CDI) fires in synchrony with the motor neuron that's controlling wing movement; it fires as the wings close. The CDI's firing then inhibits the auditory neurons responsible for processing sound—so the cricket is deaf to the sounds it generates on the wings' downbeat. When the CDI doesn't fire, and there is no corollary discharge, incoming sounds are deemed external or non-self, and the cricket tunes in.

It's not just crickets. Similar single-neuron recordings of the mechanism of corollary discharge can be seen in nematode worms, songbirds, and even marmoset monkeys.

Within a decade of Irwin Feinberg's 1978 proposal that a fault in the brain's corollary-discharge mechanism might underlie the varied symptoms of schizophrenia, Chris Frith, a clinical psychologist who was then at the Northwick Park Hospital in Harrow, UK, developed his "comparator model" for how our sense of agency arises—the sense that makes us *feel* we are responsible for our actions. At the time, Frith argued that the disruption of this very basic aspect of our sense of self was behind the first-rank symptoms of schizophrenia: auditory verbal hallucinations, thought insertion, and delusions of control (the delusion that someone else is controlling one's actions).

While the model has morphed somewhat over the years, its essence remains the same. Say you want to move your arm. The motor cortex sends commands to the muscles in the arm. The motor cortex copies the command to other brain regions, which then use the copy to predict the sensory consequences of the arm movement. Meanwhile, the arm moves, which results in certain sensations (such as tactile, proprioceptive, or visual sensations). The "comparator" matches actual sensations with predicted sensations. If there is no mismatch, we feel that we performed the action—we *own* the action, giving us a sense of agency. A mismatch makes us feel that someone else, an external agency, is responsible.

It's easy to see the appeal of this model. It allows the brain to dampen its response to self-generated sensations (for instance, the cricket's deafness to its own chirps). It provides a mechanistic explanation for how the brain might distinguish between self and non-self, at least for motor actions. And there's evidence that this ability is hampered in people with schizophrenia.

Take tickling. It's near impossible to tickle yourself. Frith, along with Sarah-Jayne Blakemore and Daniel Wolpert, showed why. In studies of healthy people, the researchers found that a couple of brain regions were far less active when people touched their left hands themselves compared with when the experimenter touched their left hands. The brain was stifling its response to self-generated touch sensations (explaining why we can't tickle ourselves). Also, the brain region that is likely doing the stifling is the cerebellum, possibly by predicting the effects of self-generated movements.

Blakemore, Frith, and colleagues further showed that people experiencing auditory hallucinations and delusions of control felt a touch on their left hand as equally intense, ticklish, and pleasant regardless of whether they themselves or the experimenter did the touching. In other words, many people with schizophrenia can tickle themselves. This suggests an inability to tell apart self-generated actions from non-self actions.

There's more evidence. Judith Ford and Daniel Mathalon, at the San Francisco VA Medical Center and the University of California, San Francisco, have shown that healthy people, just like crickets, can dampen down their response to self-generated sounds. Brain EEG signals in healthy people, just prior to them uttering a sound, show a synchrony that is suggestive of a copy of the command to move the vocal cords being sent to the auditory cortex. And then, an EEG signal called N1, indicative of auditory cortex activity, is damped down about 100 milliseconds after the healthy person makes the sound. This is possible evidence that the predicted sound has been compared with the actual sound, causing the external sound to be tagged as self-generated and thus ignored. N1, however, is not suppressed when the sound is external, which indicates that the person can hear it.

But this mechanism seems to be impaired in people with schizophrenia. This is evidence of a possible disruption of the copy mechanism. For them the N1 signal is not suppressed to self-generated sounds, which means that the patients are hearing their own vocalizations in the same way they would hear external sounds (Sass suggests that this is a kind of hyperreflexivity—a propensity to take as an external object that which would usually be only tacitly experienced, and therefore be the very medium of selfhood). It is no great leap to think such disruptions of the comparator mechanism could blur the boundaries between self and non-self in schizophrenia.

● ● ○

At this point it's worth getting subtler about what exactly might be going awry in schizophrenia. When I move my hands, I have two feelings: a sense of owning my hands and a sense of agency that makes me feel that *I* am moving my hands. We saw in the previous chapter how BIID can be attributed to the loss of sense of ownership of body parts. While there is some evidence that schizophrenia results in somewhat perturbed feelings of body ownership, stronger evidence implicates an impaired sense of agency.

In 2008, cognitive neurologist Matthis Synofzik of the University of Tübingen, Germany; philosopher Gottfried Vosgerau of the Heinrich-Heine University in Düsseldorf, Germany; and their colleagues got even more picky. They argued that one's sense of agency should be subdivided into a nonconceptual (nonthinking and instinctive) *feeling* of agency and a more cognitive *judgment* of agency. Synofzik's team says that while the *feeling* of agency relies on copies of motor signals and comparators that match predictions with actual sensory feedback, the *judgment* of agency depends on a cognitive analysis of the environment and our beliefs about it, which is called postdiction. "If you are alone in a room and something falls down from the table, your world knowledge will tell you that things do not fall by themselves, so you conclude that it must have been you, even if you don't have a sensory motor feeling of having done anything," Vosgerau told me during a phone conversation.

Of course, it's all happening in the blink of an eye, so to speak. Nonetheless, it's possible to tease apart these mechanisms. Researchers have shown that people with schizophrenia have a disturbed *feeling* of agency, and to compensate they tend to rely more on their *judgment* of agency, which depends on external factors such as visual feedback. This means that, on the experiential level, they are likely to experience themselves almost as if from outside themselves, again manifesting a kind of hyperreflexivity and an absence of a more basic sense of existing. This could also explain the split-second delay Sophie said she experiences between moving her hands and feeling that she initiated the action—a delay that makes her question whether her hands are her own.

None of this negates the comparator model. In fact, Synofzik and colleagues acknowledge that their results "support the notion of a dysfunction of the comparator mechanism in schizophrenia." Indeed, it's because of this dysfunction that people with schizophrenia have to rely more heavily on their judgments about the external environment to augment their sense of agency.

So, if a person with schizophrenia picks up a television remote and switches on the TV, he might not feel that he initiated that action. The television nonetheless comes on, so the patient infers someone else made him do it. In Laurie's case, she didn't feel like she cut herself after an evening of watching fireworks on Bonfire Night. "Although it appears to be my decision, it was not my decision, or my volition, to do such a thing," she told me. "So [there is a] loss of agency, yes."

Given she *knew* she didn't decide to cut herself, the alternative was obvious: somebody else must be responsible. "I think it's a natural search for meaning. This is happening to me, so I want an explanation, just like any other human being would do," she said. "So, then you have an enemy, a conspiracy." Paranoia is often the outcome.

In a way, the comparator model and its variants help us understand why a person with schizophrenia may feel like his actions are controlled by an external agency and how it might lead to paranoia. Or why the sounds one utters may seem like they were spoken by someone else. But what if no one is speaking, not even you, and yet you hear voices?

● ● ○

Judith Ford has spent the past decade and a half thinking about auditory verbal hallucinations (AVHs), science-speak for hearing voices. In the late 1990s, Ford made the switch from studying aging and Alzheimer's to studying these voices. In the beginning, she'd analyze data collected by other researchers and write papers. "I was raising small children, and it worked for me," she said. But soon she realized she had to talk to her patients, pay attention to their individual experiences. It's these discussions that highlighted the nuances of what she was trying to study. For example, one of her patients told her that before he started taking an antipsychotic drug called Zyprexa, the devil talked to him. Once he was on Zyprexa, God began talking to him. He was still hearing voices, but they had gone from being negative to positive.

Such insights have informed Ford's work. Healthy people hear voices too, but they tend largely to be positive, and the individuals have some semblance of control over the voices. Not so for people with schizophrenia, about 75 percent of whom hear voices. The voices sound real and are often spoken by "specific non-self voices." They are usually negative, inciting violence toward oneself (as in Laurie's case) or others, at times leading to suicide or even homicide.

This phenomenon of voices inciting violence against others is captured vividly by Anne Deveson, an Australian writer and documentary filmmaker, in her book *Tell Me I'm Here*, in which she chronicles her teenage son Jonathan's devastating schizophrenia and the toll it took on Jonathan, Deveson, and her family. Suffering severely, Jonathan had long taken to disappearing from home, and reappearing suddenly. He could become violent. In one harrowing section of her book, Deveson describes the scene when she and Jonathan's probation officer, Brenda (who had just been summoned), confront him:

When Brenda arrived Jonathan was lying on the big couch that faced the sea. He was nodding to himself as if he were listening to voices, but he did not speak aloud. We asked if he were hearing voices. Jonathan looked suspiciously at both of us, then said, "No voices." He said something else but his voice trailed away. Brenda leaned forward and said she could not hear him.

"I said only Anne's voice," he shouted.

"Where's Anne's voice?"

"Plotting against me. Inside my head."

"Jonathan, I'm not plotting against you. And I'm not inside your head. I'm here."

He looked at me, his eyes darting everywhere, and still that racing energy which seemed to fill the whole room, bouncing off the ceiling and the walls, jangling my own energy, so that I felt I was receiving an electric shock.

"God has said that I should kill you Anne, and Brenda too if she doesn't shut up."

He stalked out of the room, waving his arms. A few seconds later he returned, looked at us both, muttered something and left again. This time he didn't return.

There is something deeply unsatisfactory about trying to find mechanistic explanations for Jonathan's complex auditory hallucinations. But science has to begin somewhere. One theory tries to explain such auditory verbal hallucinations as misperceived inner speech or inner speech that somehow is not tagged as belonging to the self. We are all familiar with inner speech—it's our internal monologue, externally inaudible, sometimes clear enough to ourselves, even if it does not have an auditory quality, and at other times experienced in a more implicit manner (in all likelihood, you are experiencing it as you read this sentence). But Ford argues that auditory verbal hallucinations are not like the willed kind of inner speech, they are more like unbidden thoughts (the stuff of daydreaming or mind wandering). The question then is: how can mind wandering turn into AVHs?

Ralph Hoffman of Yale University and his team have found that in people with schizophrenia, there is hyperconnectivity between language areas of the brain and the putamen, a deep-brain region that has been linked to the conscious perception of sound. Hoffman argues that this hyperconnectivity is what allows activity in the language areas to enter one's consciousness as voices.

To dig deeper into this problem, Ford and her colleagues looked at a network of brain regions in 186 patients with schizophrenia who heard voices, each of whom was scanned while resting inside an fMRI scanner for six minutes. This data was compared with data from 176 healthy volunteers. In the case of healthy volunteers, mind wandering while at rest showed activity in a network of the following brain regions: *the medial prefrontal cortex (MPFC)*, which is the most active region when your brain is at rest and is part of the default mode network, and is also strongly correlated to self-referential mental activity (it lights up, so to speak, when you detach from focusing on an external task and are thinking about yourself); *Broca's area*, in the frontal part of the left hemisphere, which is implicated in speech production; *the putamen*, which as we just saw is involved in the conscious perception of speech; *the amygdala*, which is deep inside the temporal lobe and is involved in the fear and threat response; *the parahippocampal gyrus*, which is known to be more active when someone becomes suspicious; and *the auditory cortex*, which, as the name suggests, is involved in hearing.

But in patients who hear voices, the scans showed that all these brain regions are hyperconnected: the MPFC is hyperconnected to Broca's area, the putamen, and the auditory cortex; and the putamen is hyperconnected to the auditory cortex. All of this, Ford and colleagues speculate, could be turning the idle thoughts of healthy mind wandering into the pathological, audible voices of schizophrenia. And what about the negative tone of these voices? It could be that the hyperconnected amygdala and parahippocampal gyrus—both of which are normally involved in the fear response—increase the levels of fear, uncertainty, and suspicion associated with these voices.

There's one final piece to this puzzle. Why do these voices feel as if they belong to someone else? As we saw earlier, Ford's work with EEG signals has shown that the efference copy/corollary discharge mechanism is disrupted in people with schizophrenia. And in these fMRI studies, the researchers found that in patients who hear voices, Broca's area and the auditory cortex are less well connected—possibly corroding the pathway for the efference copy to reach the auditory cortex. So the voices, which for healthy people would at least seem to be their own, sound foreign in schizophrenia.

"The raw material of auditory verbal hallucinations, I maintain, is not [willed] inner speech, but unbidden thoughts," said Ford. And later, in an email, she expressed it more personally, referring to her deceased mother. "In fact, when my mind is wandering and unbidden thoughts are becoming conscious, I can hear the tonality, prosody, and affect of my mother's voice telling me 'you are trying to do too much, dear'. I do not think she is speaking to me from her grave," she wrote. "But, if I were psychotic, I might."

In a psychotic person, then, a hyperconnected network might be turning unbidden thoughts into audible voices, voices that have a dark tone about them. A disturbed sense of agency makes these voices seem to belong to others.

At the heart of this malfunctioning system is what's increasingly being referred to as the "predictive brain." Generating the sense of agency is one example of how the brain's predictive mechanisms work to create our sense of self. This idea is gaining ground. Could the entire brain be a prediction machine, generating not just the sense of agency but even emotional feeling states that give us our sense of being embodied? As we'll see in the coming chapters, neuroscientists are applying such ideas to explain depersonalization disorder and even something as complex as autism.

● ● ○

It is one thing to experimentally study the disturbed feeling of agency, and another thing to explain the full panoply of symptoms that this supposedly begets in schizophrenia. This baffling and often terrifying diversity is captured by psychologist and therapist Lauren Slater in her book *Welcome to My Country*. This is how she describes her first meeting with a group of six chronic schizophrenia patients:

> There is Tran, nicknamed Moxi, a small, cocoa-colored Vietnamese who came to this country after the war, and who bows to invisible Buddhas all day in the corridors. There is Joseph, with a mangy beard, a green-and-khaki combat helmet he puts on the pillow next to him when he sleeps. Charles is forty-two years old and dying of AIDS. Lenny once stood naked in Harvard Yard and recited poetry. Robert believes fruits none of us can see are exploding all around him. And then there is Oscar, 366 pounds,

and claiming constant blow jobs from such diverse females as the Queen of England and Chrissy, the Shih Tzu dog next door.

When confronted with such patients, many find it hard to accept that a mere disturbance of the sense of agency could be responsible for all the devastating symptoms of schizophrenia, as Chris Frith hypothesized when he first put forth his comparator model. Soon after Frith's proposal, it became clear that the feeling of having others' thoughts in your own head was hard to explain using his model. Today, even he admits that his model fails to account for thought insertion. Synofzik, Vosgerau, and colleagues think that their model, which splits the sense of agency into a feeling and a judgment, does a better job of explaining thought insertion; in their view, an impaired judgment of agency leads to the feeling of having alien thoughts in one's head.

Others are not convinced either way. Louis Sass, for instance, while he agrees that the neurobiology of a disturbed sense of agency is consistent with the idea that schizophrenia is a basic disturbance of the self, questions whether the impaired brain mechanisms are the *cause* of schizophrenia. He calls that a "materialist" assumption. What if you could alter the way healthy people relate to their own experience—maybe through intense introspection or meditation—and show that their brains also undergo the same kinds of neurobiological changes as those seen in people with schizophrenia? That would show that such changes are correlated, not causative.

Ralph Hoffman has similar things to say about schizophrenia. Yes, scientists (including him) have found neural-system dysfunction and gross anatomical changes in the brains of many schizophrenic patients. But are these changes the *cause* of schizophrenia, or are the observed changes the result of "oftentimes profound withdrawal from social interactions, work and school" that can pre-date the onset of schizophrenia? "So, if you take somebody during their late adolescence and early adulthood and have them go into that stage of withdrawal and have them continue that way for years . . . what's going to happen to brain systems in the absence of cognitive enrichment and task engagement?" says Hoffman. "I hypothesize that at least some of what we end up crediting to 'neurodegenerative processes' may be the downstream consequences of the state of withdrawal that these people go into."

Hoffman is struck by the fact that psychotic symptoms are a form of interaction of the self with others. He hypothesizes that in individuals deprived of meaningful social interaction, psychotic experiences flood in to fill the void. "What happens is that in the absence of being linked into a set of real-world meanings and role specifications and places to really engage, the person becomes increasingly preoccupied with the psychotic experience that then causes further withdrawal," says Hoffman. "The internally generated experience becomes more and more prominent and it can happen relatively quickly. It kind of challenges the old breakdowns of mind, body, and brain."

It also challenges any notion of there being only a one-way interaction between the extended narrative self and the more basic self-as-subject (Sass and Parnas's *ipseity*, or Zahavi's *minimal self*): it's not necessary that only the perturbations of the self-as-subject lead to disturbances of the narrative self; the effects could flow the other way too. Also, schizophrenia is telling us that the sense of agency—which goes unquestioned when it's working well—is an aspect of the self, a constituent of the self-as-object. Even in the direst cases of schizophrenia, there is a self-as-subject that is experiencing psychosis. Who or what is that "I"?

For someone with schizophrenia, all of this philosophizing is cold comfort. And for insightful, high-functioning adults like Laurie and Sophie, an awareness of their condition can be a burden. For instance, if you are able to sometimes see through your psychosis, but not at other times, how do you tell when you are being psychotic? "One doesn't lose all the biographical, semantic, perceptual, and body memory of the past, of what the world should feel like," said Sophie. "It's that disconnect between what things are like now and what your entire life before [psychosis] was like."

There's even an official term for this quandary: "double bookkeeping"—a concept from early-twentieth-century psychiatry that has been elaborated in recent years by Sass, often in dialogue with Sophie and other persons who have experienced schizophrenic psychosis. Patients are forced to deal with two, even multiple, versions of reality. "You are almost constantly forced to make decisions that other people aren't going to make. What are you going to prioritize, which possible version of reality are you going to privilege?" said Sophie. "What are you going to act on?" Confronted with such dilemmas, patients often lapse into total inaction. This phenomenon hints at the power of the narrative self: without a coherent story about oneself, one seems unable to act; it seems that we need our narrative to function.

Laurie, too, is well aware that the voices in her head, her paranoia, the messages she thinks she's receiving from outside, are all, in some sense, a product of her altered self. "But that insight is a paradox. Without the insight you fear the external; with the insight you fear yourself," she told me. "Without insight, you think everybody else is after you, or someone else is [responsible for your actions], but with insight, you realize it's all in your head. That's also scary, so you can't win."

5

I AM AS IF A DREAM

THE ROLE OF EMOTIONS IN THE MAKING OF THE SELF

How far do our feelings take their colour from the dive underground? I mean, what is the reality of any feeling?

—Virginia Woolf

Forever I shall be a stranger to myself.

—Albert Camus

When I told Nicholas I'd come visit him, I had badly underestimated Canada's vastness. It took me six hours to fly from San Francisco to Boston, then came a ten-hour-long road trip from Boston to Saint John, New Brunswick (thanks to two wrong turns in remote parts of Maine), a three-hour smooth ferry ride across the Bay of Fundy to Nova Scotia, and then another hour and a bit to drive to Kingston, where twenty-three-year-old Nicholas lived with his fiancée and their daughter, a toddler at the time.

Once the ferry reached Nova Scotia, I drove to the village of Kingston on a highway that ran the length of the valley carved by the Annapolis River. It was late June, early summer. The countryside was lush green. The land had shed its springtime sparkle and was bursting with rude health. Purple lupines lined the road. I reached Kingston within a few minutes of exiting the highway and found my destination—a split-level white apartment building behind a 24/7 convenience store. Nicholas was expecting me and came out to greet me.

I, however, wasn't expecting to see someone with as many tattoos as Nicholas. Those covering his neck made it seem like he was wearing a T-shirt under his light blue dress shirt. And despite his rolled-up sleeves, the skin was barely visible on his right hand and forearm. There was an intricate Koi fish on the forearm, a symbol of overcoming adversity. Farther down, on the hand, was a compass for "finding direction in life," and a diamond, because "they stand up well to pressure." "Pretty clichéd," he admitted. On his left wrist were the initials of the members of his foster family, who had taken him in as their own after he finished rehab at the age of sixteen, and where he felt he belonged for the first time in his life. "It was almost like a normal childhood," crammed into three or four years.

When he closed his fists and held them together, below the knuckle on each finger was a letter; the words they formed read SINK and SWIM. Another cliché, he said. "It's very basic, and it's what it comes down to for me, especially with this disorder. I'm either going to continue fighting to get better, and hopefully be in remission again at some point. Or . . . I don't know . . . sink . . . not fight anymore." He struggled to elaborate on what "sink" would mean. "Maybe it's suicide. Maybe it is simply not trying anymore."

● ● ○

Nicholas's earliest memories are from when his sister was born. He was four years old at the time. The parents and the two siblings were all living together. But it was hardly a happy family. Both his parents were addicts. His father, who mainly did roofing and other construction work, was an alcoholic. His mother, who stayed home, was a heavy drinker and addicted to opiates like Oxycontin and Dilaudid. Nicholas had already spent a year in foster care by the time he was three years old, only to be returned to his parents by a judge. A year later, his sister was born. But nothing changed. Their parents continued doing drugs, drinking alcohol, fighting, and even disappearing for days on end, leaving the kids with other family members.

Then, their mother left their father. "And my father passed out for a couple of days, in a semiconscious state, from what I remember," Nicholas told me. The foster-care agency came to take the kids away. They found little Nicholas trying to take care of his sister ("She was so, so little"). He recalled that he had tried to make her cereal. When the foster agency workers arrived, he was standing on a chair at the kitchen sink, trying to wash dishes. The children spent the next few years in foster homes.

Around the time Nicholas was nine, the kids returned to their mother, who had since remarried. The situation with their stepfather wasn't any better. The couple had moved on to harder drugs, like crack cocaine. They screamed and fought all the time, drugs fueling their paranoid delusions, as each accused the other of hiding pills or smoking the last of the crack cocaine. Dramatic scenes were routine, but some are etched in Nicholas's memory more than others. Once, a few hours past midnight, Nicholas and his sister woke up to shouting from their parents' bedroom. The siblings walked over. Nicholas told his sister to wait at the door while he went in. He watched his stepfather push his mother against an old television set, which fell down. Another time, his stepfather chased his mother with a machete. She ran into the bedroom and locked herself in. "I don't know if he had any intent to hurt her or just scare her," said Nicholas. His stepfather ended up burying the machete into the door of the linen closet.

Their house was in a tranquil, upscale neighborhood in Bridgewater, Nova Scotia, where people kept the lawns of their "quarter-million-dollar" homes neatly mowed. The house Nicholas lived in was the odd one out. It was smaller than the rest. The family had been put there by a social program that provided housing to those with low income. Most of the time, the shenanigans inside the house were hidden to the outside world: Nicholas's parents were careful to cover the windows with blankets.

In addition to the neglect from his early years, Nicholas was now dealing with an abusive mother and stepfather. Much of it was verbal and emotional abuse. " 'You are a fucking idiot, can't you do anything right? What's wrong with you?' Stuff like that . . . very strong words," Nicholas said. Very occasionally his stepfather would beat him. "Luckily that was rare. I am really glad for that, although sometimes you wonder which lasts longer, the physical or the emotional." Was he sexually abused? I asked. "No," he said. "I'm thankful for that as well."

By the time he was about ten or eleven, Nicholas had started having brief, transient episodes of dissociation—maybe lasting about ten seconds—episodes that happened at random, sometimes when he was on the school bus, sometimes while singing the national anthem at school. "I'd describe it as feeling absolutely disconnected from the physical body altogether," said Nicholas. "It almost renders you unable to communicate, to do anything for those ten seconds."

It came to a head when Nicholas was about twelve years old. He and his sister were in their bedroom when they heard a scream from the kitchen. It was their aunt. She and their mother might have been smoking crack; Nicholas doesn't remember. What he does remember is the sight of his mother convulsing on the kitchen floor. She was having a seizure. She had hit her head on a cupboard handle as she fell and was bleeding. She was foaming at the mouth. Nicholas's stepfather came running and turned her onto her side, so she wouldn't choke on her vomit. For Nicholas, it was a pivotal moment. "I remember taking about three or four steps [toward] her, and everything completely changed," Nicholas said. "It was like I went from a normal waking state to a dream state immediately. Everything became very foggy. Everything looked unfamiliar, out of place."

For the next four years, Nicholas lived in this foglike state: where everything—the things around him and his own body and self—felt unreal. An extended, disturbing dream.

● ● ○

In a book published in 1845, the German psychiatrist Wilhelm Griesinger wrote about a letter from a patient to the eminent French psychiatrist Jean-Étienne Dominique Esquirol.

> Even though I am surrounded by all that can render life happy and agreeable, in me the faculty of enjoyment and sensation is wanting or have become physical impossibilities. In everything, even in the most tender caresses of my children, I find only bitterness, I cover them with kisses, but there is something between their lips and mine; and this horrid something is between me and the enjoyments of life. My existence is incomplete. . . . Each of my senses, each part of my proper self is

> as if it were separated from me and can no longer afford me any sensation. . . . I no longer experience the internal feeling of the air when I breath[e]. . . . My eyes see and my spirit perceives, but the sensation of what I see is completely absent.

Esquirol himself had written about other such patients and their experiences: "An abyss, they say, separates them from the external world, I hear, I see, I touch . . . but I am not as I formerly was. Objects do not come to me, they do not identify themselves with my being; a thick cloud, a veil changes the hue and aspect of objects."

What these patients were describing would today be called *depersonalization*. The word itself entered the psychiatric lexicon in the 1890s, when French psychologist Ludovic Dugas used it to describe "a state in which the feelings or sensations which normally accompany mental activity seem absent from the self." Dugas chanced upon the word in the diaries of Swiss philosopher Henri-Frédéric Amiel. In his book *Journal Intime*, which was published after his death, Amiel wrote: "I find myself regarding existence as though from beyond the tomb, from another world; all is strange to me; I am, as it were, outside my own body and individuality; I am *depersonalized*, detached, cut adrift. Is this madness?"

The twentieth-century German psychiatrist Karl Jaspers gave a particularly clear description of what might be happening when a person is feeling depersonalized. Everything that manifests itself in our mind, "whether perception, bodily sensation, memory, idea, thought or feeling carries *this particular aspect of 'being mine'* of having an 'I' quality, of 'personally belonging,' of it being one's own doing. This has been termed *personalization*. . . . If these psychic manifestations are accompanied by the awareness that they are not mine, but are alien, automatic, independent, arriving from elsewhere, they are called *depersonalization*."

There are some who argue that transient depersonalization is an evolutionary adaptation to extreme danger. In the mid-1970s, Russell Noyes Jr. and Roy Kletti of the University of Iowa College of Medicine interviewed sixty-one people who had responded to an ad in the student newspaper asking for "accounts of subjective experiences during moments of life-threatening danger." A typical response was that of a twenty-four-year-old man who recounted the moments during which his Volkswagen skidded while cornering a "rain-slicked curve" into the opposite lane with oncoming traffic. "As the car was spinning I had a relaxed kind of feeling like being stoned on 'pot' or something," the man reported. "I gave no consideration to the danger, it just didn't exist. I had a sensation of floating. It was almost like stepping out of reality. I seemed to step out of this world, where you feel the sensation of your body in the seat and the air you breathe, into some other state."

Based on their interviews, Noyes and Kletti concluded: "The interpretation of depersonalization as a defense against the threat of extreme danger or its associated anxiety seems inescapable. . . . Thus, in the face of life-threatening danger, persons become observers of that which is taking place and effectively remove themselves from danger. Detachment appears to be a major adaptive mechanism which, in the depersonalized state, is seen in bold relief."

If depersonalization is indeed an evolutionary adaptation, it makes sense that we would all have the intrinsic ability to enter such a state in which we become strangers to ourselves. Given that such neurobiological mechanisms exist, it also makes sense that some of us would slip into it more easily than others. Call this predisposition (nature). Then environment (nurture) would play its part in tipping some over.

For instance, an abusive childhood and the resulting trauma could lead to depersonalization, as likely happened with Nicholas. Drugs can do it too.

● ● ○

Sarah is a slim, small woman—wired, energetic, in her early thirties, and already running an online start-up in New York. We met in her office and walked over to a nearby café to discuss an experience that she was still trying to understand. "What the hell just happened?" she asked rhetorically.

Three weeks before we met, she had been visiting a friend in the East Village. It was Saturday evening. Her friend liked to smoke pot recreationally. She joined in too, though she could count on her fingers the number of times she had smoked pot before. Sunday morning her friend suggested they try Adderall. Usually prescribed as a stimulant to treat attention deficit hyperactivity disorder (ADHD), Adderall has become a drug of choice among those working in the intense environments of tech start-ups as a way to focus and enhance performance. Sarah's friend, an undergraduate, was taking Adderall to study harder and longer. Sarah had never tried it before. They split a pill. But their day of stimulation wasn't done yet. That evening they had some pot again, smoked a hookah, and had a drink.

Sarah woke up Monday morning in a bit of a haze. "You feel hazy after you have a fun weekend, so I didn't think much of it," she told me.

Tuesday morning she went for a six a.m. yoga class. She was still feeling hazy, disconnected. "It felt like I was dreaming," she said. "I was questioning whether I was alive." Still, she kept her work appointments, made phone calls, had meetings.

By Wednesday morning, things hadn't improved. In fact, they felt worse. She started crying. "Shit, did I die? Am I looking at my life now?" she recalled thinking. "I kept saying, 'I need to get out of this. I am in something I don't understand.' I was panicking. The thing that kept bothering me was that I didn't believe anything."

She doubted herself and the reality of her environment. She kept feeling she was in a dream. She got on the train to go to midtown for a meeting, and found herself spooked. She looked at her fellow passengers and wondered, "Are they here? Am I here?"

Wednesday evening—it was the night of a new moon—Sarah went to pray at her liberal Jewish synagogue, something she had grown up doing. On the way, she stopped at a taco stand and ordered tacos, fully convinced that the tacos wouldn't show up. Standing on the New York City sidewalk, she said to herself, "They are not going to show up; there's no way I ordered these tacos. I'm going to wake up, and it's going to be a dream." But the tacos were soon handed to her, of course. "I literally said, 'Oh my God, I'm alive. Phew!' "

Sarah was likely doing what psychologists call "reality testing"—there was an awareness of objective reality, but her subjective sense of it was messed up, and she kept trying to reconcile the two. The momentary confirmation of getting real tacos, however, was of little help. Throughout the evening prayer, she remained anxious, her head churning with an internal dialogue seeking confirmation that she existed—because she felt disconnected from herself. When the prayer ended, she avoided eye contact with anyone and rushed back home. She started crying again. "I'm dead. I have died. I have absolutely died," she thought.

Thursday, she called her nurse practitioner, who said, "Look, there is no way the drugs are [still] in your body." The nurse advised fluids and rest. It didn't help. Friday morning, a friend—whom I know well also—came to meet Sarah. As soon as our mutual friend saw her, Sarah blurted out, "I'm afraid that I just died." They sat for a few moments together before deciding to go to a nearby ER—a short, six-block walk away. Sarah sobbed all the way to the ER, and continued sobbing in the ER. "A really strange sob," she recalled.

She continued asking her friend, "Are you sure I'm alive, am I here? Are you here right now?" Her friend would reassure her, "yes, yes, we are alive." The ER doctor put it all down to a reaction to the drugs (even though he said it was odd that it had continued for so long after Sarah had taken them), prescribed her Ativan for anxiety, and sent her home.

But Sarah's feelings of unreality continued. Life continued to feel dreamlike—a bit untethered, as if she were floating; she struggled to describe her state in everyday parlance. She's not sure why it eventually all cleared up about three weeks later. The passage of time, or maybe the massage she got from a massage therapist, who was also an "energy-healer." "Listen, I am fucked up right now," Sarah told the masseuse, and explained how she was feeling. The woman replied reassuringly, "Oh, happens to me too. You have to learn to manage it." She told Sarah that energy was flowing in and out of her body from the top of her head, and she had to stop the flow using a twisting motion of her hand, as if closing an imaginary bottle. Sarah was ready to believe her. The masseuse also gave her some less esoteric advice: "Do yoga; you need to feel back in your body. Stomp your feet, do whatever you need to feel back in your body, and you'll be fine."

Sarah did feel a tiny bit better when she left the masseuse. It was time to do more yoga, which she had been doing regularly anyway each morning for the past six years. The day after the massage, she did something extreme even by her standards—two hour-long sessions of yoga back-to-back. "I did one class and I was going to leave, and one of the teachers said, 'You should stay.' I was like, 'I'm still dreamy, what else am I going to do?' so I stayed. I felt a little bit better."

The recovery continued and by the time we met about three weeks after the onset of her experience, she was fine.

● ● ○

"You need to feel back in your body. . . . Stomp your feet, do whatever you need to feel back in your body."

There's wisdom in that seemingly woolly advice. Neuroscientists may cringe at counsel from a New Age healer, but if you unpack much of what neuroscientists themselves are saying about the self, it's clear that feeling embodied—being in and of our body—is a key aspect of who we are.

One of the strongest advocates of the notion that the body underpins the self is neuroscientist Antonio Damasio. In his latest book, *Self Comes to Mind*, he writes, "Of the ideas advanced in the book, none is more central than the notion that the body is a foundation of the conscious mind." Damasio believes that a conscious mind needs the appearance of a self. His framework begins with the idea of mental images forming in the brain. By this, he does not mean images in our conscious awareness. Rather, he means patterns of activity of neural circuits. Depending on the pattern of activity, a given neural circuit within the brain can be in one of many, many states. Each state, writes Damasio, is the equivalent of a mental image. A succession of such mental images is the mind. At this stage, Damasio is not talking about anyone or anything being conscious of the mind. This is not hard to accept. It's well known that much of our brain's activity is subconscious—we are not, and will likely never be, aware of it.

The brain's basic job is to take care of the organism as the organism eats, drinks, moves, and sleeps. An organism can survive only if its internal biochemistry remains within acceptable limits. These are so-called homeostatic limits, and the process that keeps the body within these limits is called homeostasis (a term defined by American physiologist Walter Cannon). The exact homeostatic process differs from organism to organism. For instance, cold-blooded animals (such as reptiles) take on the temperature of their surroundings—so in order to be active, they have to seek warmth—whereas the bodies of warm-blooded animals (just about all mammals and birds) need to be at a near-constant internal temperature, which entails eating a lot when cold to generate the necessary energy or sweating to lose excess heat. The brain performs its duties of maintaining homeostasis admirably. But the brain is part of the organism too, not something distinct sitting outside the body like a puppeteer pulling strings. In order to do its job, the brain has to keep track of—create representations or maps of—what's happening in the body, in the environment outside, and within itself. The patterns of activity of neural circuits are these maps. These maps are the contents of (the still unconscious) mind.

The next step in his hypothetical framework is the appearance of what Damasio calls the *protoself,* "which foreshadows the self to be." The protoself is made up of mental images of the more stable aspects of our body (such as the state of the viscera). Damasio implicates the upper brain stem as the region most responsible for creating maps and generating such images. The structures in the upper brain stem are inextricably linked to the body parts they map—a tight two-way interaction that is "broken only by brain disease or death."

The other key function of the protoself is the generation of *primordial feelings,* which "provide a direct experience of one's own living body, wordless, unadorned, and connected to nothing but sheer existence." Primordial feelings, in Damasio's framework, "reflect the current state of the body."

The next layer, structured on top of the protoself, is the *core self.* The core self is constituted of representations in the brain that capture the relationship between the protoself and its interaction with an object. The representations also capture changes to the protoself and the resultant primordial feelings. For example, if the protoself were to encounter a snake, the core self would be the representation of this interaction and the representation would include the changes to the body state (a state of utter fear for someone like me).

Damasio thinks that with the appearance of the core self in our evolutionary history, the self as we know it entered the picture. The core self is the first intimation of subjectivity (note, however, that Damasio does not satisfactorily explain how neural activity turns into subjectivity—no surprise, since that's the stuff of the hard problem of consciousness). The core self is living in the moment—it is a sequence of mental images of interactions of the protoself's dealings with an object, and the ensuing modifications to the protoself and primordial feelings. If all we had was a core self, and many animals likely do, then all we'd be aware of are these moments of subjective experience.

It's when the brain evolved further and developed autobiographical memory that the next stage came about—the *autobiographical self.* Damasio hypothesizes that there is brain circuitry that is capable of grouping together autobiographical memories into an object (one can think of this object as a story), letting that object interact with and modify the protoself, which then produces a moment of subjectivity. But this time the subjective experience is not just of the body, but of a more complex entity, the person. The autobiographical self would be a rapid sequence of such moments of subjectivity. This fully formed self would be the basis of one's personality.

Regardless of whether one agrees entirely with Damasio's framework, there's widespread agreement in neuroscience about the body's central role in giving rise to the self. This role is manifest in emotions and feelings. In Damasio's view, the self begins with primordial feelings—body states represented in the upper brain stem, the insular cortex, and the somatosensory cortices—which form the building blocks for more complex emotions and feelings.

Damasio also proposes an "as-if body loop"—roughly translated, the ability of the brain to simulate body states. Why would a brain want to do this? Because, at times, simulating an anticipated state can speed up the brain's capacity to control the body's physiological state and thus save energy. It can make the brain more efficient and effective. It's not unlike the idea of the brain generating an efference copy of a motor command, using it to predict the sensory consequences, and being prepared for it.

If the body and primordial feelings form the basis for our sense of self, depersonalization—given that it involves disembodiment and the numbing of emotional feelings—can be viewed as a fundamental impairment of self-awareness. Mauricio Sierra and Anthony David of the Depersonalization Research Unit at King's College London write that "the condition manifests as a pervasive disruption of self-awareness at its most basic, preverbal level (i.e. what it feels like to be an entity, to exist)."

Depersonalization will mostly induce some or all of these experiences: "(1) feelings of disembodiment, which refers to the sense of detachment or disconnection from the body; (2) subjective emotional numbing, an inability to experience emotions and empathy; (3) anomalous subjective recall, a lack of ownership when remembering personal information or imagining things; and (4) derealization, an experience of feeling estranged or alienated from surroundings."

● ● ○

People with depersonalization find it hard to put their experience into words, resorting mainly to metaphors. "The actual disconnect itself is very challenging to describe," Nicholas told me. "It feels like your physical body is not you."

It almost sounds like an out-of-body experience, but it's not. There are some key differences between the experience of feeling disembodied in the context of depersonalization and the disembodiment of out-of-body experiences (which we'll examine in detail in a later chapter). Depersonalization does not usually involve a shift in perspective, in which the observing self somehow ends up outside the body. This shift in perspective is common in out-of-body experiences, however, and is suggestive of different neural mechanisms that work to keep the self anchored in the body. In depersonalization, one is (usually) still located within the body, but the vividness of being embodied is compromised.

Nicholas was about twelve when he became chronically depersonalized—an episode that lasted four years. "The most terrifying thing about having it at that age was that I didn't have any support," he told me. No parents, no teachers, no friends that he could speak to. His sister was too young to understand.

It didn't help that they didn't have a stable home. Their mother and stepfather were eventually arrested on charges of drug abuse and child neglect. Nicholas and his sister were again living in group homes, or with foster parents. Around the time he turned thirteen or fourteen (the time line is hazy in his mind, something he attributes to depersonalization), Nicholas was reintroduced to his biological father. His mother had filled him with hatred for his father. Nicholas wanted to see what he was really like. By then, his father had spent four years in jail, was tattooed—not the colorful ones like Nicholas now has, but the dark, bluish-black ones common among prisoners—and had bulked up in prison. "He was some two-hundred-odd pounds of muscle," said Nicholas. "He looked like a body builder."

Moving in with his father turned out to be disastrous. His father was living with his own stepfather, and both were doing drugs. They paid no heed to Nicholas. In fact, Nicholas's stepgrandfather (for want of a better word) would buy him alcohol. The young Nicholas, barely fourteen, slipped headlong into a world of liquor, weed, and other drugs. He was soon intravenously injecting himself with morphine. "That was my rock bottom, as far as addiction goes," he said. Meanwhile, his foster family—whom he had run away from to live with his father—informed the authorities that Nicholas was in serious danger.

Nicholas's now fiancée, Jasmine, recalled the day he was whisked away to rehab. Jasmine was with her friends in downtown Liverpool, Nova Scotia, when a friend ran up yelling, "Nick was arrested!" Jasmine had met Nicholas just the week before. Suddenly he was off to jail. But it wasn't jail he was going to: the Department of Community Services was taking him to rehab. At first he was sent to a secure care facility in Truro, where he detoxed for a month. Then he spent nine months in a rehab center in Sussex, New Brunswick. Sometime during the rehab, he went into remission—his depersonalization disappeared.

When he finished rehab, a young couple in their thirties—Tammy and Dave—took him in. I spoke to Tammy about how she and her husband were charmed by the well-spoken and likeable teenager.

"When we met Nick I'd say we both fell in love with him," she told me. "By the time we encountered any difficult times with him, we both loved him like our own." Nicholas had serious difficulties, however. He was ridden with anxiety; he could only be alone while sleeping. He was terrified of large buildings. "He didn't go to Walmart, for example," said Tammy. "There was something about the way it looked and felt that meant he couldn't get out of the car at Walmart." Tammy also noticed something amiss about the way Nicholas felt about people. "Especially with new relationships, he spoke of not feeling what he thought other people must feel. And yet, things still seemed to hurt him." She also doesn't remember him exulting in anything. "He certainly has been happy in his life, but not elated."

The couple took all that in stride, which meant the world to Nicholas. "They treated me as if I was their own child. It was huge for me. I lived there for three to four years, and in those three to four years I learned responsibility, accountability . . . all kinds of things I hadn't been raised to learn," said Nicholas. He learned to drive, got his license, went back to school, and got his diploma. That's also when he started getting his tattoos—something that his foster parents weren't terribly keen on, but nonetheless they got a kick out of seeing Nicholas tattooing their initials on his wrist.

The remission from depersonalization was itself anticlimactic. Life returned to normal. No firecrackers went off. "Depersonalization becomes a very, very faint memory in the background," said Nicholas. He quit smoking (he had sworn off drugs, except for one relapse after rehab), exercised a lot, and life was relatively good. "When I look back at my life, those three years are the zenith of my life," he told me.

During his remission, Nicholas sought a lawyer's help to see his file at the Department of Community Services. In the file he saw the list of disorders he had been diagnosed with: "Depersonalization disorder, OCD, generalized anxiety disorder, and another one called oppositional-defiant disorder."

His remission wouldn't last. One day, while at work in a call center, Nicholas drank a large can of an energy drink, which was loaded with caffeine and taurine. It triggered a massive panic attack. The attacks came frequently thereafter. "They started to get worse and worse and more frequent, and very, very intense," recalled Nicholas. "[I'd be] thinking that I'm dying at the time of the panic attack." Worse yet, the depersonalization returned, something he had forgotten about.

"Even the simplest things feel strange when you have depersonalization," he said. "You become so hyperaware of things. Opening and closing your hands, or moving your arms when you walk, or even walking itself, all those things become very strange, because you do not feel it's you doing them. It feels like you are sending commands to somebody else to do it for you." (This is reminiscent of what Sophie and Laurie said about their experiences with schizophrenia. Many people with schizophrenia show signs of depersonalization in their prodromal phase, before progressing to full-blown schizophrenia.)

Meanwhile, Nicholas started dating Jasmine. The relationship was hard at first. Jasmine would point out that Nicholas didn't seem to care, or be emotionally invested; he seemed distant, preoccupied. Slowly, Nicholas explained that it wasn't him; it was his depersonalization that was the problem. He felt emotionally numb toward everything.

The numbness remained even after they got engaged. "It's like she is not actually my fiancée. I know that she is and I know that I love her, but it doesn't feel like she is somebody I know. It's almost like not recognizing somebody but you do. It's odd," Nicholas told me. "I have talked to other people about this too who had the same thing. They know they love the person they are with, and they are aware of that, but it feels that the person is a stranger. You don't have a full connection."

And then his daughter was born. Nicholas was in the delivery room helping. He watched his daughter enter the world. "I had waited so long for her to be born, and it was such a big event. I cried when she was born and I felt it. For her birth, I was tapped in," said Nicholas. "I haven't experienced [such feelings] since, but I'm so glad I did experience that. There have been so many situations, with her, and with deaths of friends, where I didn't feel it fully, but for some reason, my daughter's birth was an exception to that."

● ● ○

Emotional numbness in depersonalization is a paradox. It's clear that people who suffer are unable to feel intensely—as is evident from Nicholas's descriptions—yet they are distressed and panicked, which are also feeling states.

Nick Medford, a neuropsychiatrist at Brighton and Sussex Medical School in England, recalled a woman patient of his who exemplified this paradox. The family living next door to the patient had just suffered a horrible tragedy: their young child had been killed in a terrible accident. "She knew that the appropriate things to say were 'that's terrible, I'm so sorry, that's awful,' but she said she didn't feel anything about it," Medford told me. "But then she felt disturbed by the fact that she didn't feel anything."

Another patient told him, "I don't have any emotions—it makes me so unhappy."

"It's sort of contradictory," said Medford. "If you unpack that, I think what people are describing is that they have a lot of internal emotional distress or turmoil, but they don't seem to have emotional reactivity to external things."

It's very clear that people with depersonalization are experiencing muted emotions, an altered sense of one's body, and an altered sense of reality. Something is amiss in the body-brain system that generates feelings of body states. Sufferers are also prone to self-rumination—that is, giving excessive thought to their altered state, and potentially greatly reducing the attention they pay to the external world (recall Steven Laureys's finding about external and internal awareness networks, and how they are inversely correlated—one works at the expense of the other). Self-rumination may also "contribute to the sense that the world has become somehow distant and unreal."

Jeff Abugel, an author of two books on depersonalization, and the person who introduced me to Nicholas, is someone who knows about such obsession. He has experienced transient episodes of depersonalization since his late teens. During these episodes, "virtually everything that constituted my life mentally had kind of disappeared. The only thing that was left was this nonstop sensation of trying to figure out what was wrong with me," he said. "My whole existence became just pondering: what's wrong with me, why do I feel this way, what's going on?"

So, while the feelings of distress and unhappiness seem to be the result of an obsessive focus on the sense of strangeness, the strangeness has its basis in the way emotional feeling states are generated—and how they underpin the sense of self.

Medford and his colleagues have studied the emotional response of patients while they lay inside a scanner. If a person with an intact emotional system is shown emotionally positive, neutral, or negative images, the scanner shows brain activations appropriate to each type of stimulus. One of the brain regions that is activated when viewing emotionally salient images is the insula. Activity in the insula is correlated with "every conceivable kind of feeling," writes Damasio in *Self Comes to Mind*, "from those that are associated with emotions to those that correspond to any shade of pleasure or pain, induced by a wide range of stimuli: hearing music one likes or hates; viewing pictures one loves, including erotic material, or pictures that cause disgust; drinking wine; having sex; being high on drugs; being low on drugs and experiencing withdrawal; and so forth" (recall the case of the sixty five-year-old woman with dementia who suffered from Cotard's syndrome: she had bilateral insular atrophy, potentially messing with her bodily feeling states). In depersonalization, Medford's team found that there is distinctly less activity in the left anterior insula while viewing aversive images when compared with healthy controls. "The emotional circuitry, emotional responses, seem to be switched off somehow," Medford told me.

The switch lies elsewhere in the brain. Another brain region that has been regularly implicated in depersonalization is the ventrolateral prefrontal cortex (VLPFC)—an area of the brain that's involved in top-down control of emotions. Medford's study (one of the largest ever done, with fourteen depersonalization patients) found that the VLPFC was overactive in these patients when compared with controls. An overactive VLPFC might be suppressing emotional responses in depersonalization.

The team took the study one step further. While there are no known medications for depersonalization, some people have reported improvements when they have taken lamotrigine, an anticonvulsant prescribed for epilepsy. Ten of the fourteen patients in Medford's study took lamotrigine for four to eight months, after which they agreed to be scanned again. Some patients reported that their condition had improved, while in others there was no change. Those whose symptoms had abated showed increased activity in the left anterior insula and decreased activity in the VLPFC when compared to the scans from before they began taking lamotrigine and when compared to the scans of those who were not feeling better despite the pharmacotherapy. "Whereas the people that hadn't improved at all, they were still very flat in terms of neural responses," said Medford, of the activity in the insula.

The left anterior insula is involved in integrating sensations from both inside the body (interoceptive) and outside (exteroceptive), and is thought to be crucial for creating a subjective sense of our own body and indeed for the sense of self. Neuroanatomist Bud Craig, who has done seminal work to understand the neuroanatomy of the insula, argues that it provides the neural substrate for the "sentient self." Antonio Damasio begs to differ (arguing that the brain stem too has an important role to play in representing body states).

While the VLPFC in people with depersonalization can be said to be "switching off" the left anterior insula, it's not under conscious control. "It's not a willed thing," said Medford. "It's just happening. Things are being switched off."

If so, this switching off should become apparent in how autonomic nervous system responses—which are not under conscious control—operate in people with depersonalization. And in fact, that's exactly what researchers have seen: if you measure skin conductance of the hand (an autonomic response) in reaction to unpleasant stimuli, people with depersonalization show very little activity. "When you have a patient with depersonalization wired up to measure skin conductance, you are constantly checking to see if the thing's actually [connected]," said Medford. "Because you just get this flat line, which is not what you normally see."

So, given that depersonalization makes people feel like strangers to themselves, and given that their ability to experience emotions is muted, what does this tell us about the self? "It's telling you about the primacy of physical sensations and internal sensations" in the making of the self, Medford said. "This Damasio-type idea that feelings are built up from somatosensory information."

Damasio himself built on ideas that date back to the late 1880s, when William James challenged existing beliefs about emotions and feelings by asking: when you see a bear, do you run because you feel fear, or do you feel fear because you run?

● ● ○

"Common sense says, we lose our fortune, are sorry and weep; we meet a bear, are frightened and run; we are insulted by a rival, are angry and strike," wrote William James in 1884, in a classic paper called "What Is an Emotion?" James suggested that this is the wrong sequence of events, however, and put forth his new hypothesis: "We feel sorry because we cry, angry because we strike, afraid because we tremble, and not that we cry, strike, or tremble, because we are sorry, angry, or fearful, as the case may be."

What are the modern neuroscientific definitions of emotion and feeling? An *emotion* is the physiological state of the body in response to stimuli. The state includes not just aspects like heart rate and blood pressure but also the motor movements of the body (freezing or fleeing in response to a threat, for instance). An emotion also includes the nature of cognition in that state (for example, is your thinking razor-sharp or dull?). A *feeling* is the subjective perception of this emotional state of the brain-body complex.

When James wrote his paper, the commonsense view was that we feel first and then act, leading to the various behaviors that characterize a given emotion. So, if you saw a snake, and if you are someone who is afraid of snakes, then according to this rather intuitive view, you'd first feel the fear, and the feeling would then drive you to take action, which could be to either flee the scene or freeze in terror.

James argued that we had misunderstood the relationship between emotions and feelings: it was the other way around. And even though his use of the word "emotion" wasn't quite in line with modern neuroscience, his larger point was well taken. We emote first and then feel the emotion.

At the same time, and independently, Danish physiologist Carl Lange proposed a near-identical idea, and so it came to be called the James-Lange theory. But Walter Cannon, who coined many words and phrases in use today in the field of physiology, including "homeostasis" and "fight or flight" to describe an organism's response to threat, did not take kindly to the James-Lange theory. He pointed out, for example, that when people were injected with epinephrine (otherwise known as adrenaline), it produced many physiological changes that were similar to a natural emotional state of arousal, but the people did not necessarily *feel* the artificially induced emotional state. In other words, changes in body states did not result in the expected feelings. This seemed to go against James's notion that feelings follow emotion.

Cannon's formidable reputation ensured that the James-Lange theory languished until the 1960s. Then, elegant experiments by Stanley Schachter and Jerome Singer resurrected the James-Lange theory, but with modifications. The scientists recruited subjects to study the supposed effects on vision of a fictious drug called Suproxin. In reality, the subjects were injected either with epinephrine or a placebo (saline solution). When injecting, the experimenter did one of three things: (1) spoke about the exact side effects of the injection (pounding heart, shaking hands, and a warm, flushed face); (2) gave the subject a false account of possible side effects (itching, numbness of feet, headaches); or (3) said nothing at all.

Of course, the subjects all thought they had been given Suproxin. The experiment involved a further twist. When a subject was waiting for the drug to take effect, a "stooge" who supposedly had also been injected with Suproxin came and sat in the room and proceeded to vigorously act out either a state of euphoria or anger. The idea was to see if the context influenced what the subject would feel.

The experiments were revealing. The essence of the results was that feelings (of anger or euphoria) seemed to depend not just on the physiological state of the body but also on the cognitive context—which led the subject to "appraise" the emotional state of the body. The cognitive interpretation of the physiological change caused by the injection—influenced by the stooge's behavior and what a subject had been told about the drug's possible side effects—played a part in what the subject eventually felt or experienced. "Cognitive factors appear to be indispensable elements in any formulation of emotion," wrote Schachter and Singer.

A spate of follow-up experiments were somewhat inconclusive; some even failed to replicate Schachter and Singer's results. But the general idea persisted—that feelings are the result of an appraisal of an emotional state, an appraisal that is influenced by one's context.

Interestingly, experiments done with beta-blockers—which interfere with beta receptors all over the body and negate the effects of epinephrine or adrenaline—were more revealing. Beta-blockers essentially inhibit the flow of information to the central nervous system about the body's state of arousal, resulting in reduced levels of anxiety. "Removing cues from visceral arousal does reduce the intensity of some emotional experiences," writes psychologist James Laird in his book *Feelings: The Perception of Self.*

Laird attributes the somewhat inconclusive results of experiments that followed Schachter and Singer's to one factor: the experiments did not take into account people's differing abilities to respond to cues about the state of their own body. They did not account for varying abilities of interoception, the perception of sensations from within the body.

Still, such two-factor theories (integration of bottom-up information about the state of the body and a top-down appraisal of it) continue to be favored among emotion researchers.

Damasio and colleagues have, over the past two decades, argued that this interaction between emotion and cognition cuts both ways: while cognitive context influences our appraisal of the emotional state of the body and the consequent feelings, cognition itself can be influenced by the emotional state.

Anil Seth, co-director of the Sackler Centre for Consciousness Science at the University of Sussex, thinks that there is a better way to think about the brain basis of emotion—and take away the split between cognition and physiology. His view hews to the increasingly popular idea that the brain is a prediction machine, and that especially when it comes to external signals, what we perceive is the brain's best guess as to the cause of those signals. Seth has extended this idea to how the brain deals with internal signals from the body and argues that it has consequences for understanding disorders like depersonalization and the bodily aspects of our sense of self.

● ● ○

Nicholas knows only too well the importance of feeling connected to one's body. "I have never been so aware of what your core is as a person until I got depersonalized and felt disconnected with my physical body," he said. "Honestly—and I'm not just saying this because I have it—I think one of the scariest things a human being can endure is the feeling of separation between your physical body and your mind, but being completely cognizant of it the whole time. It's like being eaten alive."

He has discussed his ongoing depersonalization with his physician. After Nicholas signed a release form, I spoke with her over the phone, and she confirmed that she had been treating Nicholas for anxiety, which had improved, but he had not shown much improvement in his depersonalization. She had referred him to a neurologist to rule out temporal-lobe epilepsy (which can sometimes cause depersonalization)—but the waiting times for specialist care in Nova Scotia being what they are, Nicholas was still dealing with his illness on his own.

I asked Nicholas about the guitar I saw in his living room. He's learning the instrument. He'd prefer to play drums, he told me, but he is not allowed to play them in the apartment they live in. He's waiting to move into a house where he can begin drumming again. It brings him relief from depersonalization. "Drumming requires all your attention," said Nicholas. "When you are using all four limbs, it takes so much attention that it allows you to experience some relief."

I was reminded of a patient Nick Medford had mentioned. The patient had been a good amateur tennis player in London but had stopped playing tennis because of depersonalization. "The only meaningful thing that I managed to do for him was to persuade him to start playing tennis again," Medford told me. "When he was running around the tennis court, completely immersed in the flow of that, [the depersonalization] would lift. It would come back again, unfortunately, but it was nevertheless quite significant for him, because it proved to him that it wasn't fixed; it was malleable."

Nicholas also pointed out that while drumming alleviated his symptoms, the relief was nevertheless transient. The moment he became aware that he was feeling better, the depersonalization returned. "It's a paradox; you think about the fact that you are feeling better, and you start feeling depersonalized again," he said.

Could the complex phenomenology of depersonalization be the outcome of a predictive brain gone wrong?

● ● ○

Think for a moment about the quandary of being a brain. It has to deduce the nature of physical reality based on constantly changing streams of sensory inputs, which are themselves modulated by the moving body. How does the brain turn stimuli into perception?

The nineteenth-century German physiologist Hermann von Helmholtz suggested that the brain solves this problem of perception by making inferences about the causes of sensations. It acts, in modern neuroscience parlance, as a Bayesian inference engine.

The word "Bayesian" comes from the Bayes theorem, which was developed by Thomas Bayes, an eighteenth-century English mathematician and clergyman. The theorem links the conditional probability of an event P occurring given Q has occurred to the conditional probability of Q occurring given P has occurred. Bayes's theorem is used widely in so-called Bayesian networks, which are at the heart of many modern artificial-intelligence systems. For instance, AI systems in medicine use Bayesian networks to diagnose diseases: given a set of symptoms and test results, the system calculates the probabilities for various causes, offering up as a diagnosis the one with the highest likelihood of causing the symptoms. As another example, consider the case where a patient tests positive for Ebola, but the test itself is accurate only nine out of ten times. What is the probability that the person has Ebola? Our naïve intuition will want to rely on just the test and say the probability is 0.9. Our intuition would be wrong. The probability depends where the person is being tested. If he or she is in a country where Ebola is endemic, the conditional probability that the person is infected is higher than if the person is tested in a country where there is zero incidence of the disease.

A Bayesian brain, in theory, functions similarly. It computes the most likely cause of sensory inputs based on prior beliefs it has about the probable causes of such inputs. As alluded to earlier, the brain's best guess as to the cause of the sensations rises up, so to speak, as perception. Of course, this is an ongoing process. The brain uses internal models of the body and the world to predict the expected sensory input. Any difference between the expected signals and the actual signals constitutes a "prediction error." The brain uses these error signals to update its bank of prior beliefs, so that it can predict (and thus perceive) more accurately when similar signals come again.

Such "predictive coding" models, which use Bayesian inference, have mainly been applied to explain exteroception: making sense of external sensations coming from outside the body. Interoception also involves perception, but in this case it's making sense of the signals from within the body. The brain needs to know the body's state in order to determine if the body has moved away from its biochemical comfort zone and whether it needs to initiate actions to get the body back to a physiological state that's optimal for survival. Anil Seth's argument is that predictive coding should hold true for making sense of internal bodily signals too. "It is just another process of perception," he told me.

Seth's argument has a bearing on emotions and feelings. The two-factor models of emotion always involve the integration of information coming into the brain via incoming nerves, generating a snapshot of the physiological state of the body, which then is subject to cognitive interpretation and subsequently gives rise to the feeling of an emotion. With predictive coding, the division between cognition and physiology is done away with. At no place in the brain is there incoming information being integrated to create a perception. Rather, it's prediction all the way—what you perceive and what you *feel* is always the brain's guess as to the cause of the signals. This sophisticated idea evolved from the same thinking that gave us corollary discharges and comparators, models that have been used to explain how the brain gives rise to a sense of agency and suggest their impairment may lead to the symptoms of schizophrenia. Turns out that predictive coding can be applied to just about anything and everything the brain does.

Seth argues that there would be many levels of predictive coding in the brain. The lowest level would be predicting the causes of the incoming sensory signals from the body. The prediction would form an input, or incoming signal, to the next level in the brain and so on. It's a hierarchical model that is well suited to how the brain is structured. "The subjective emotion we feel is the brain's best predictive guess that explains the [incoming] interoceptive information at a whole bunch of hierarchical levels," said Seth. "It's not just cognition looking down at physiology and interpreting it."

Predictive coding is a new way of thinking about the brain and how it carries out its remit, with the potential to explain interoception and exteroception, emotions and feelings, and indeed explain how psychopathologies emerge when this predictive mechanism goes wrong. As we'll see in the next chapter, the model is even being applied to explain something as complex and varied in its symptoms as autism. But there is "also a danger," said Seth: "You explain everything, you explain nothing."

The main criticism of the predictive-processing approach is that there is no direct evidence for it. But there is evidence consistent with predictive coding. For instance, the work on the mechanisms of corollary discharges/comparators is considered circumstantial evidence for predictive brain mechanisms. There's also evidence that the insular cortex, tucked beneath the temporal and frontal lobes, is the brain region most likely involved in doing comparisons between top-down predictions of expected interoceptive signals and incoming signals that contain information about prediction errors.

Accepting such circumstantial evidence for now, the big question for psychopathology is what happens when there are errors in prediction. Errors are an indication that whatever the brain is modeling at any given level is not quite right. There are two options at this point: the brain can either update its model and bring it in line with the sensory inputs; or it can initiate some action to push the body to the desired state. The latter mechanism would be the basis for homeostasis (say you wade into the frigid waters off San Francisco and stay in a little too long—your internal body temperature will start dropping below limits that are deemed acceptable by the brain's model of your viscera, and you will feel the urge to get back to warmer environs).

In this way of thinking, the brain's function is to minimize prediction errors. And this has consequences for our sense of self. Take the signals that are coming from one's own body. In Seth's thinking, when the brain's internal models are accurate, and there is a good match between the predicted and actual interoceptive signals, you get a sense that you are embodied, that your body and emotions belong to you. The match between predicted and actual signals implicitly tags the body and associated emotions as "self," whereas a mismatch would be akin to tagging it "non-self." So, the vividness of emotional feeling states and the sense that they are *mine* depend on the brain making accurate interoceptive predictions and minimizing the corresponding prediction errors.

But what if there are persistent and ongoing prediction errors, either because of faulty internal models of the body in the brain or because something goes wrong in the neural circuitry that compares and generates errors (it's intriguing that the insula, which is key to models of predictive coding and interoceptive inference, is one of the brain regions implicated in depersonalization)?

Seth speculates—and stresses that it's just speculation—that this would then lead to dissociation, a sense of unreality about one's own body and emotions, feeling disembodied, feeling estranged from oneself. It's as if the predictive brain's best hypothesis for the source of interoceptive signals—given persistent errors—is that the signals do not belong to self, but rather to non-self.

As with everything we have encountered so far, depersonalization too does not destroy the "I." There is still the subjectivity—the self-as-subject—that is aware of being estranged from other aspects of the self, in this case the vivid emotions and feelings that give us our sense of being embodied. So, while no one is saying that our emotions and feelings are not integral to our sense of self, it's nonetheless intriguing from a philosophical perspective that they don't *constitute* the self-as-subject; the "I" stands apart, watching, observing.

● ● ○

I watch Nicholas with his thirteen-month-old daughter. He cuddles her in his arms, kisses her, almost as if he's not going to let her endure the lost childhood that he did.

Is he in touch with his father? I ask him.

"I don't talk to my father anymore," he says. "Stopped talking to him about a year ago."

"Anything trigger that?" I ask.

He hesitates.

"You don't have to talk about it if it makes you uncomfortable," I say.

But Nicholas does talk. When he lived with his father nearly a decade ago, his father had been a muscular, tattooed man, "very hard-looking in the face . . . like a criminal." But then he went west, to Alberta, and got into harder drugs. By the time he came back to Nova Scotia, he had dropped seventy pounds and was now a small-framed man, with a weathered face and an alcoholic's nose scarred with broken blood vessels, "an extremely, extremely messed-up person."

In October 2011, things got worse. The police in Bridgewater received a 911 call from Nicholas's father. They arrived to find one of Nicholas's close friends, a young man of twenty-two, dead in Nicholas's father's apartment. (It was later determined that the young man died from a methadone overdose. Nicholas's father would be charged with prescription drug trafficking, but the case was dropped for lack of evidence.) Soon after that tragedy, Nicholas's father got into a relationship with the young man's girlfriend, who was just eighteen.

"You can almost hear Jerry Springer in the background," quips Nicholas.

The saga had an even more sordid end. In March 2012, the young woman was found unresponsive in an apartment and she too died, in a hospital in Halifax two days later—allegedly due to a prescription drug overdose.

Nicholas's father did end up in jail eventually, for drunk driving.

Nicholas's mother suffers from auditory hallucinations and paranoid delusions (possibly the result of years and years of drug abuse). She was blond, thin, and pretty when she was young. Now she's disproportionately overweight, possibly due to her medication, with extra pounds on her face and midriff, and looks older than her age, says Nicholas. He meets her occasionally. "It's not much of a mother-son relationship," he says. "We never really had that."

Does he feel sad it's turned out this way with his parents? I ask him.

"I do, definitely," he says. "At the same time, with depersonalization, I feel disconnected from it. It's odd—I feel sad about it, [but] it doesn't feel like they are my emotions. It almost feels like I'm feeling sad about somebody else's life story."

"You have been through a lot," I say.

"Yes, I have," says Nicholas. "It's been one hell of a life. I wish it felt more like my life, and not like I'm observing somebody else's life."

"Even though it's been so harsh?"

"Oh, yeah," he says. "I wish I could come to terms with it. It's hard to come to terms with something when it doesn't feel like it fully happened to you."

We talk of this and that. Of the big lizard—an Australian bearded dragon—that he keeps in an aquarium. Of San Francisco and the

Golden Gate Bridge, from where I have come. Nicholas and his fiancée are enamored by the idea of California. I point out they are living in a spectacularly beautiful place themselves. "You can't appreciate it as much if you have been here your whole life," says Jasmine.

She could be speaking about depersonalization and the estranged self. Those of us who inhabit our bodies seamlessly and without disruption and feel and own our vivid emotions may not value what we have. You can't appreciate the self as much if you have been intimately connected to it your whole life.

6

THE SELF'S BABY STEPS

WHAT AUTISM TELLS US ABOUT THE DEVELOPING SELF

Autists are the ultimate square pegs, and the problem with pounding a square peg into a round hole is not that the hammering is hard work. It's that you're destroying the peg.

—Paul Collins, *Not Even Wrong: A Father's Journey into the Lost History of Autism*

I myself am opaque, for some reason. Their eyes cannot see me. Yes, that's it: The world is autistic with respect to me.

—Anne Nesbet, *The Cabinet of Earths*

James Fahey was thirty-four when he was diagnosed as being an Asperger. When I referred to his "Asperger's syndrome" in our first correspondence, James, without taking any offense, gently chided me: "*I am an Asperger*, pure and simple—I do not have or suffer from any artificially constructed syndrome, disorder, disease, or flaw."

His diagnosis came over two sessions with a psychologist. During the second session, his younger sister joined him, to provide the psychologist some perspective on their childhoods together. It was then that James realized for the very first time that some of his perceptions of those early years differed from his sister's. It came as a shock.

For instance, his sister recalled clearly that James had never wanted to play with her, even if she was alone and he was by himself too. She wanted company, and he didn't recognize that. "I never thought that she might get lonely," said James. "I don't get lonely; why would my sister get lonely? It just never occurred to me." James would rather read by himself; even his choice of books (usually historical accounts of wars—the Napoleonic wars, for instance) set him apart from his sister. He was interested in facts and statistics. She would read children's fiction. "I can't stand fiction," James told me. "I can't see the reason why I would ever want to read it."

James grew up in the western suburbs of Melbourne, Australia. His parents weren't particularly physical in their affection, which was fine by him. "I won't say it was a cold or neglectful upbringing; my sister came out fine," he said. James was uncomfortable being touched or hugged. His extended family on his mother's side didn't pose problems: they were mostly farmers, living out in the countryside. "They stood away from you; they didn't need to have a close-proximity conversation," recalled James. "They didn't want to hug you all the time."

Interactions with his father's family, however, were an ordeal. Most Sundays, James's family would drive over to his paternal grandmother's place, where she'd be waiting, at times with his great-aunts. They'd want to hug and kiss him. "At the time, I couldn't think of anything worse than someone planting a big, sloppy, slimy kiss on me," said James. "And I didn't like being hugged. I felt trapped, locked in a cage."

He was well into his twenties when he began realizing that who he was and people's notions of how he should live his life were incompatible. Having a girlfriend, for instance. Social pressure made him try, but he found it difficult. "How many girls are going to accept a guy who wants to go and be by himself all week?" he told me. "That's not much of a relationship." And the fact that he didn't like touching didn't help matters either. Any intimacy had to be structured. "Want some intimate moments, that's fine, if it's organized at a particular point in time. Go to bed, you do your thing, that's fine. Don't touch me again."

It's not that James is against relationships, he just doesn't want the intimate kind. "I'm not sad or depressed about this, not in the slightest, given how much I enjoy and need solitude. Platonic relationships suit me fine. They provide most of the positives, without most of the negatives."

Even in less-demanding social interactions, James is aware that he can get things wrong. "Maybe laughing inappropriately, schadenfreude moments," he said. Once, while watching a movie with friends, James began giggling about something he noticed in the background. It had nothing to do with the main story line. But his mirth drew attention. The movie was *Schindler's List*. His friends thought he was amused by Jews being rounded up. But James had been fixated on something minuscule, unrelated. In a way, it's how he sees the world around him. To explain, he referred to *Schindler's List* again—there's a scene in the mostly black-and-white movie in which the main character, Oskar Schindler, sees a girl in a red coat. Later, we see a red coat among a pile of bodies being carted away. Our attention is drawn to that splash of color. "That's kind of a fair analogy for how I see things in the world," James said. "Some things are highlighted to me. I follow them around. I don't know why I assign this degree of interest to particular things."

Well before he got his official diagnosis (a term he hates—"makes me sound like I have a problem"—even though he values much of what psychiatrists do), James had thought long and hard about why he was different. Socializing made him anxious, depressed. Even today he's anxious around most people. His introspective nature, however, helped him make sense of who he is and eventually shed the burden of others' expectations. "I mistook my own self for what they believed I should do," he said. "They just couldn't understand my perspective."

The kind of traits that James's introspective insight laid bare—his desire to be alone, his acknowledged difficulties in forming intimate social bonds, anxiety in social situations, and traits from his childhood, such as an aversion to being hugged and kissed—came under scrutiny in the early 1940s, when the word "autism" was used for the very first time in medical literature.

● ● ○

"Autism" comes from the Greek word *autos*, meaning "self." In 1916, Swiss psychiatrist Paul Eugen Bleuler coined the term to describe a symptom of schizophrenia—a symptom that has been described as "the narrowing of relationships to people and to the outside world, a narrowing so extreme that it seemed to exclude everything except the person's own self."

Then, in 1943, Leo Kanner, an Austrian-born psychiatrist and director of the Children's Psychiatric Service at Johns Hopkins Hospital in Baltimore, wrote a landmark paper in which he used the word "autism" to describe the syndrome with which it's now associated. It was an elaborate paper, detailing keen observations of eleven children:

> The . . . fundamental disorder is the children's *inability to relate themselves* in the ordinary way to people and situations from the beginning of life. Their parents referred to them as having always been "self-sufficient"; "like in a shell"; "happiest when left alone"; "acting as if people weren't there"; "perfectly oblivious to everything about him"; "giving the impression of silent wisdom"; "failing to develop the usual amount of social awareness"; "acting almost as if hypnotized."

Kanner differentiated his notion of autism from Bleuler's account of the symptoms of schizophrenia, in which the withdrawal from normal social relationships begins late in childhood, or even in adulthood. About this new syndrome, Kanner wrote, "There is from the start an *extreme autistic aloneness* that, whenever possible, disregards, ignores, shuts out anything that comes to the child from the outside."

In a notable illustration of the emergence of an idea whose time had come, Hans Asperger, an Austrian pediatrician living in Vienna, independently published a similar paper of case studies a year later, also using the word "autism" to characterize the condition.

It wasn't until 1980, however, that the *Diagnostics and Statistical Manual of Mental Disorders* (DSM), the oft-quoted, much-maligned publication of the American Psychiatric Association, first included autism as a diagnostic category, albeit with the unfortunate phrasing: "infantile autism." This was renamed in 1987 as autistic disorder. But the difficulty in defining autism became evident in the DSM-IV (1994): autistic disorder was now placed alongside several more subtypes, including Asperger's disorder and pervasive developmental disorder not otherwise specified (PDD-NOS). In 2013, the fifth edition of the manual (DSM-5) reversed course, rolling back Asperger's and PDD-NOS into the umbrella term *autism spectrum disorder.*

Through all this categorization and recategorization, Kanner's original insight remains: "We must, then, assume that these children have come into the world with *innate inability* to form the usual, biologically provided affective contact with people, just as other children come into the world with innate physical and intellectual handicaps," he wrote in 1943. Or, as he put it, the children had "innate autistic disturbances of affective contact," where "affective" refers to the emotional aspect of our being.

● ● ○

The function of a self, by definition, is to help an organism discern the boundary between itself and others. Is an infant born with this ability? Or does a developing baby form a self in stages? There's certainly good evidence that by about eighteen to twenty-one months of age, babies begin to distinguish between themselves and others while speaking. To the horror of some parents, infants start screaming "mine" for things they fancy, a clear indication of the ability to *explicitly* reference one's self using language. Around the same time, a child is able to recognize his or her reflection in a mirror or in photographs. One of the questions facing developmental psychologists is this: does the child develop a sense of self primarily through social interactions and the use of language, or is there an underlying, more innate, and implicit self?

William James distinguished between the implicit and the explicit, calling the former "I" and the latter "Me." As psychologist Philippe Rochat puts it, "The 'Me' corresponds to the self that is identified, recalled, and talked about. It is the conceptual self that emerges with language and which entails explicit recognition or representation. It is beyond the grasp of infants, who by definition are preverbal, not yet expressing themselves within the conventions of a shared symbol system. On the other hand, there is the self that is basically implicit, not depending on any conscious identification or recognition."

In 1991, Ulric Neisser, often called the father of cognitive psychology, further split James's "I" into an *ecological self* (an implicit sense of the body that babies develop in relation to their physical environment) and an *interpersonal self* (an implicit sense of the social self that emerges via interactions with other people).

The ecological self is evident in babies. For example, babies have a rooting reflex: touch them on the cheek and they will turn their heads toward the touch. Rochat has shown that even newborns, within twenty-four hours of birth, are three times more likely to show rooting behavior when someone else touches their cheek than when they accidently touch their own cheek. This shows that babies are aware of their own bodies, maybe even as agents of their own actions (recall from the chapter on schizophrenia that we are unable to tickle ourselves because the efference copy/corollary discharge mechanism dampens down the sensations; we implicitly know we initiated the tickling).

According to Rochat, babies develop an implicit social self because adults are constantly mirroring the facial expressions and emotions of babies (how often have you seen a mother sympathize with a crying infant by making a sad face herself and saying, "My poor baby"?). Babies in turn are themselves imitation machines. It's this constant mirroring that helps the baby develop a preverbal, prelinguistic social self—one that is honed by interpersonal interactions.

With language comes the ability to form and express the explicit self. As they grow further, children develop another intriguing ability: the ability to peer inside the minds of others, so to speak. In children with autism, however, this ability is impaired, leading to what Kanner called an innate inability to relate socially to others. For researchers who are also thinking of autism in the context of the self, questions loom: is the problem of social-relatedness linked to the child's own developing self and, if so, how?

● ● ○

Susan and Roy recalled the first time they realized that something was amiss with the way their son, Alex, was acting around other people. It was his second birthday. Relatives and friends had come to their home to celebrate. Alex was the center of attention, except he didn't seem to want it to be so. "He was running out into the street, onto the sidewalk," Roy told me. "I had to drag him back in. I clearly remember most of the guests looking a little confused."

But even before that incident, it was obvious that the little boy was hypersensitive to sounds and touch. "You couldn't hug him a lot," said Susan. He'd only wear soft cotton clothes, with the tags removed, because the tags irritated him. Loud sounds were anathema: he'd often cover his ears and show clear distress. Even his food had to be bland and without too much texture (nothing chewy or crunchy), or he'd gag over it. All this led to an initial diagnosis of sensory processing disorder. There were hints, however, that Alex was dealing with more than just tactile and auditory hypersensitivity.

Even before his second birthday (which was the first big wake-up call for Roy and Susan), a health-care worker had come to do a wellness check on their son. Alex had a collection of dozens and dozens of toy cars. "The minute she came in and saw that he had lined them up in a straight line, it's like some red flag went up for her," said Roy. If anyone disturbed Alex's particular arrangement of cars, he'd get upset.

However, he wasn't proprietary about his toys. Most children start developing a sense of ownership of their toys, but Alex didn't display this tendency. "He just never had that," said Susan. "Anybody could come play with his toys; anyone could take something away." Alex was also diagnosed with expressive speech delay. While his language skills and comprehension were age-appropriate, he didn't talk much about his emotions and feelings (he didn't say, for example, "I feel happy, or mad, or sad," said Susan) in the same way that most kids his age would.

Meanwhile, teachers at his preschool were noticing Alex's anxiety. For instance, when the kids would be sitting in a circle, awaiting their turn to answer a question or talk about something, Alex would start getting nervous as his turn neared. He'd bite his nails and rock back and forth. Eventually, Alex was diagnosed with PDD-NOS, a subtype that has since been subsumed into the larger autism spectrum disorder in DSM-5.

By the time he entered elementary school, his preference for aloneness manifested itself on the playground. "He was a very solitary person. The whole playground would be full of screaming kids playing with each other, and he'd be off by himself, not playing with anybody, not interested in playing with anybody," said Roy. "And that was a major thing, because all the teachers would notice it, and it'd come up in the yearly discussions we had."

After years of various kinds of therapies (speech, physical, and occupational), which all parents of autistic children are well aware of, Alex has come a long way. He's sharp academically, and has developed a keen interest in sports. "He's a popular kid; he's very affable, very kind, he'll never bully anybody," said Susan. "Because he is bright, often kids come to him and ask him for help with math and things. He shows no attitude. He's generally very, very well liked whichever class he's in."

Yet, said Roy, "He doesn't have any close friends, even now."

Alex seems to regard everyone the same. "If you are talking to kids, they'll say stuff like, 'This kid is better than the others,' or 'This one is kind,' or 'This one is good,' or 'I like so-and-so more than I like this one,'" said Susan. She doesn't see this in Alex. "He's not very reflective about what he's done, or what others have done. He's not discerning in that sense. Pretty much everything is fine with him."

For Alex, and many kids like him, this has implications for the way they function in society as they grow up. Alex is part of a group that takes part in social language therapy. One of their field trips was to go to a shop and order something. This meant discussing scenarios over and over: asking for something, anticipating what the man behind the counter might say, replying to the man, giving money, receiving change, and so on. "Stuff that you and I would pick up by inference, and know how to handle, these kids have to be taught," said Roy.

Stuff that you and I would pick up by inference... What are we inferring? We are inferring the mental states of others; we are reading minds. By the mid-1980s, child psychologists began coming up with simple tests to determine when children begin showing an ability to peek into, in a manner of speaking, someone else's mind. This ability is called theory of mind (ToM). What can autism tell us about theory of mind? And do we need a theory of mind to understand our own minds, our own mental states? Is a theory of mind necessary for a sense of self?

● ● ○

Psychologist Alison Gopnik has a unique way of getting people to appreciate the idea of theory of mind. Imagine a room full of people. Look around. What do you see? Aren't you seeing things that are "bags of skin that are stuffed into pieces of clothing and have little dots at the top that move back and forth and a hole underneath?" Do we see people as these inanimate objects? Of course not. "That's mad, that's crazy," said Gopnik, when I visited her at her office at the University of California, Berkeley. "You never see other people that way. You see them as psychological beings."

That means seeing people as entities that have minds. We are constantly inferring what's going on in someone else's mind, to understand their behavior, intentions, and desires as well as to predict what they might do next. In a sense, we are constantly theorizing about other people's mental states. We have a *theory of mind.* It's this ability that underpins human social interactions. But is this ability something that we are born with? Or does it develop over time? In other words, is there a stage during development when a child clearly acquires this ability?

In 1983, two Austrian psychologists, Heinz Wimmer and Josef Perner, published a paper on how to test whether children have a theory of mind. Their paper began with a quote from a book on artificial intelligence:

> A travelling salesman found himself spending the night at home with his wife when one of his trips was unexpectedly cancelled. The two of them were sound asleep, when in the middle of the night there was a loud knock at the front door. The wife woke up with a start and cried out, "Oh, my God! It's my husband!" Whereupon the husband leapt out from the bed, ran across the room and jumped out the window.

We have to ask, were both husband and wife in the habit of being unfaithful? Wimmer and Perner were discussing the idea of false belief. What was the wife thinking? Why did the husband jump out of the window? Attributing a false belief to another person is being able to infer that what they are thinking does not tally with reality as you know it. For instance, did the wife think it was her husband knocking at the door, and did the husband infer that his wife was thinking such thoughts, even though he was lying next to her? While that doesn't explain why he jumped out of the window, at least it illustrates the idea of false belief. And being able to figure out someone else's false belief is a clear indication that you have an insight into their state of mind, that you have a theory of mind.

Wimmer and Perner wanted to test whether children could attribute a false belief to another person and then predict what the person would do. They carried out a series of ingenious, if somewhat involved, tests and showed that children developed this new skill—of reading minds—between the ages of four and six.

Meanwhile, at the University of London, Alan Leslie, a postdoc with developmental psychologist Uta Frith, was looking for evidence of theory of mind in another seemingly innocuous childhood activity: pretend play. Leslie argued that newborns have an innate ability to model the world around them—an accurate internal "primary" mental representation. But this ability cannot explain pretend play, such as pouring pretend tea from a toy kettle into a toy cup. Such pretend play requires a child to be able to do two things: have a primary representation of actual reality and another for the made-up world. Leslie called pretense the "beginnings of a capacity to understand cognition itself":

> It is an early symptom of the human mind's ability to characterize and manipulate its own attitudes to information. Pretending oneself is thus a special case of the ability to understand pretense in others (someone else's attitude to information). In short, pretense is an early manifestation of *theory of mind*.

And evidence was mounting that children with autism, especially those whose autism was severe, did not indulge in pretend play or fantasy (unlike, say, other children with intellectual disabilities, such as Down syndrome, who can pretend play, although they are slow to reach that stage of development relative to normal kids).

I asked Roy and Susan whether this had been true of Alex as he was growing up. "Oh, totally," said Susan. "He played with his trucks and games and all, but that constant narrative of [pretending that] 'I'm this and I'm going to be this or that'—absolutely not."

It was such evidence of lack of pretend play that led Leslie to posit that autistic children should show an impaired theory of mind, but not so children with Down syndrome. In 1985, Simon Baron-Cohen, with his PhD adviser Uta Frith and Leslie as co-supervisor, devised a simpler, more elegant version of the Wimmer and Perner experiment to test theory of mind in autistic children. It was called the Sally-Anne test, after the two dolls that were used to enact a scenario, which went like this:

> Sally has a basket. Anne has a box. Sally has a marble, which she puts into her basket, and goes out for a walk, leaving the basket behind. When Sally's away, Anne takes out the marble and puts it into her box. Sally comes back into the room and

wants to play with her marble. Where will Sally look for her marble?

If you answered "Anne's box," then you have failed the false-belief test—because you think that what's in Sally's mind is the same as what's in your mind. If you answered "Sally's basket," then you have passed—you are able to peek inside Sally's mind (she should look in her own basket, because she could not have known of Anne's duplicity).

As predicted, children with autism (with a mental age above four years) struggled with this task—they were more likely to answer "Anne's box"; whereas typically developing children and those with Down syndrome were more likely to answer correctly. This suggested that the autism involved a specific deficit of theory of mind.

By 1988, Alison Gopnik and her colleagues, working with typically developing children, showed that theory of mind and tests of false belief could illuminate something deeper about the very essence of self-awareness: that the ability to know our own mind is related to the ability to know another's mind.

They showed a group of three-to-five-year-old children a closed box of candy. However, when the children opened the box, they were surprised to find pencils inside instead of candy. The box was then closed and the children were asked a series of questions that were designed to test whether they knew that their current knowledge (or mental representation) of the contents of the box was different from what they had thought the box contained before it was opened. While the five-year-olds remembered that they had earlier harbored a false belief about the contents of the box, the three-year-olds forgot that they once had a false belief that the box contained candies, not pencils. As far as they were concerned, the box always had pencils. The experiment had tested the children's ability to recognize false beliefs, except in this case it was the child's own false belief at an earlier time.

It's intriguing that the ability to read another's mind and the ability to discern what had been in one's own mind in the past both develop between the ages of three and five. "There is a very strong correlation between what children say about the other person and what they say about their own past self," Gopnik told me.

Baron-Cohen studied children with autism using similar experiments, the results of which were published in a paper provocatively titled: "Are Autistic Children 'Behaviorists'?" The experiments essentially tested children to see if they could distinguish between the physical appearance of an object (something that looks like an egg) and the knowledge about its real nature (upon touching, the egg turned out to be made of stone, and only looked like an egg). This so-called Appearance-Reality (A-R) experiment involved seventeen autistic children, sixteen mentally handicapped children, and nineteen clinically normal children (all children had a verbal mental age of at least four years). Baron-Cohen argued that this ability to tell apart appearance from reality was a test of how aware the children were of their own mental states.

The children were first shown the object. When asked what it was, all children answered that it was an egg. Then, the children got to touch the "egg," and discovered that it was actually made of stone. Once they had the chance to examine the stone-egg carefully, they were asked two questions: "What does it look like?" and "What is it really?" The correct answers would be egg (appearance) and stone (reality), respectively.

About 80 percent of the typically developing children and children with Down syndrome passed such A-R tests. However, only about 35 percent of the children with autism were able to distinguish appearance from reality: they mainly made the error of saying that the stone-egg looked like an egg and was an egg. "This suggests that these children alone are unaware of the A-R distinction, and by implication unaware of their own mental states. These results suggest that when perceptual information contradicts one's own knowledge about the world, the autistic child is unable to separate these, and *the perceptual information overrides other representations of an object*," wrote Baron-Cohen. The autistic child's brain is unable to fully utilize prior knowledge to make sense of new information—a clue that the brain's predictive mechanisms may be impaired.

These studies are telling us something profound. Whatever the brain mechanisms that help us to read the minds of others, the same mechanisms seem to be involved in reading our own mind. How important is reading minds, or having a theory of mind, for our sense of self? "I would consider it crucial," Uta Frith told me.

But crucial for a certain aspect of our self, said Frith. As we have seen, the self can be broadly divided into two: a prereflective self-awareness (the "I," or the self-as-subject) and a reflective part (the "me," or the self-as-object). Theory of mind has to do with the self-as-object. No one is claiming that autism impairs a child's ability to be a subject of experiences.

● ● ○

If children with autism have an impaired theory of mind—and possibly difficulties reading their own minds—how does this manifest in adults with autism? Do they have trouble introspecting too?

To answer this question, Frith, along with Russell Hurlburt and Frith's former PhD student Francesca Happé (who has since become a leading autism and theory-of-mind researcher), tested adults with autism.

Their technique, pioneered by Hurlburt, required the subject to carry a beeper that would beep at random, at which point the subject would have to "freeze the contents of his awareness" and write down the details of his thoughts. Three high-functioning men diagnosed with Asperger's, who all spoke and communicated well, took part in the study. The subjects were also given a battery of false-belief tests. What emerged was an interesting correlation between how well they did on those false-belief tests and their ability to introspect.

Two of the subjects, Nelson and Robert, who did well on the false-belief tests, were best at being able to introspect and write down their inner experiences, even though these experiences were reported mainly as visual images and often lacked elements that would make up the inner experiences of normal controls (in normal controls, people whom the autism community calls "neurotypicals," the experiences included perception of inner speech and feelings).

Peter, the third subject, who did not do well on the false-belief tests, struggled with introspection as we know it. "There were no reportable inner experience[s] in any of the samples. . . . There were no images, no inner speech, no feelings, etc. that other subjects have reported," wrote the researchers.

During my conversations with Frith and Happé, I was reminded of James Fahey. I had asked him whether he was close to his sister. He surprised me with his answer. "I'm not saying this is the same for all Aspergers, but for me, I don't feel [emotions] for people. I don't have a visceral sense. . . . I don't get butterflies in my stomach. I don't get heart palpitations. I love my sister, but it's done purely at a cognitive level. I *think* love for her; I don't feel love for her."

How did his sister react when he told her this? I asked. "She took it rather well," said James. "I was surprised. I thought that this is going to sound bad to a person who is used to emotional connection with others, that this may come across as quite an affront. [But] she tried to understand. Maybe having grown up with me, she saw that my emotional connection to people was very different from what it was in herself."

This extraordinary admission reveals multiple things. Firstly, James can introspect, but not about aspects of the self that neurotypicals might take as given. His self-as-object, for instance, does not include sharply felt emotions for others, a deficit that he has overcome by developing an explicit cognitive ability (this is likely why some people with autism who initially fail false-belief tests eventually pass them). But the inability to effortlessly fathom people's emotions is taxing; he has to compensate by paying conscious attention to, say, body language and facial expressions (something neurotypicals are likely doing on autopilot). No wonder social situations continue to be a source of anxiety. "The computer in my head is working in overdrive," said James. "It's stressful. Thirty minutes of socializing can be as draining as a three-hour calculus exam."

James's situation made perfect sense to Frith. "It's just the totally automatic, innate kind of ability that is missing in autism, but that doesn't mean that they cannot get there in a conscious way, via effort and via learning," she said.

This feeds into the debate over how an individual develops a theory of mind, the ability to mentalize. There are a few schools of thought. One idea is that theory of mind is a "theory theory"—a confusing phrase that aims to say that we implicitly theorize about what is going on in someone else's mind, using unconscious cognitive processes. Another says that we simulate a scenario in our minds in order to understand another's mind; you sort of go through the motions offline, as it were, of others' actions and come to some understanding of their state of mind. Yet another view is we can directly perceive another's state of mind—an inference that happens fast, a process that's below the threshold of consciousness, and enters consciousness as direct perception.

A deficit of theory of mind might also be linked to another characteristic of people with autism—deficits in so-called executive functions, which leads to difficulties in planning a sequence of actions necessary to achieve a goal.

Susan and Roy recounted Alex's problems with day-to-day tasks. "A two-year-old child will instinctively know that if you have to go out, there are some things you have to do, like wear shoes and socks, get a jacket on. You automatically know that you need to do these things before you can step out of the door," said Susan. "For Alex you have to remind him a lot more. Every time it's a new thing. It's not something that became habitual with him. Or to know that you have to put your socks on first, then to put your shoes on, or that you have to get your underwear on before your pants."

Studies have demonstrated strong correlations between difficulties with theory of mind and executive function in children with autism. The argument is that people with autism have difficulty accessing or maybe even representing mental states that model the sequence of possible behaviors that haven't yet happened (such as putting on socks, shoes, and jacket) or the desired goal (to be fully dressed before leaving the house).

This too supports the notion that theory of mind is not just for accessing the mental states of others, it's crucial for knowing one's own mind, and hence for aspects of one's sense of self. And without insights gleaned from studying autism, we might never have begun looking for brain structures responsible for theory of mind.

● ● ○

In adults, a set of brain regions is strongly correlated with theory of mind: the temporoparietal junction (TPJ), the precuneus (PC), and the medial prefrontal cortex (MPFC). These brain regions are activated when you think about what others are thinking. MIT's Rebecca Saxe studied these brain regions in children five to eleven years old, ages when they are developing and honing their theory-of-mind abilities. Turns out the same brain regions are implicated in these children in tasks that require theory of mind. In fact, the right temporoparietal junction (rTPJ) is most strongly correlated with theory-of-mind abilities in children. Cambridge University's Michael Lombardo, working with Simon Baron-Cohen and others, has shown that the rTPJ is functionally specialized for representing mental states, and that this specialization is impaired in people with autism, so the greater the degree of impairment, the greater the difficulties in relating socially to others.

Lombardo and colleagues also implicated another brain region as a problem area in autism: the ventromedial prefrontal cortex (vMPFC). The study's subjects (an apt word in this case), both neurotypicals and individuals with autism, were asked questions designed to get them to ponder their own mental characteristics or those of the British queen: "How likely are *you* to think that keeping a diary is important?" versus "How likely is the *queen* to think that keeping a diary is important?" The neurotypical controls used the vMPFC more on tasks that referred to the self than to the queen. This was not the case for people with autism: "the ventromedial prefrontal cortex treated self and other equivalently in autism."

The same study found another intriguing link: with autism, there was reduced connectivity between the vMPFC and regions of the brain that are involved in more basic body representations, including the ventral premotor and somatosensory cortex. Couple this with the fact that the right TPJ (the other region strongly implicated in the theory-of-mind deficit, as we've seen) is also linked to body maps, and a new way of looking at autism begins to emerge. It's a perspective that some researchers are taking seriously: that autism may be due to an inability to accurately perceive one's own body and the sensory stimuli that it's receiving. This would disrupt one's sense of the bodily self, and while such an interruption would have direct consequences for processing sensations, it could also affect higher-level processes such as theory of mind.

● ● ○

There's an incident from Alex's first day in first grade that Roy cannot forget. Alex was sitting at a desk with a girl he knew from day care. The teacher gave Alex and the girl some crayons and paper and asked them to draw. The young girl drew a beautiful butterfly. "Really, really cool butterfly," said Roy. But Alex barely managed some "random stuff," said Roy. "It was way behind the standard which you expect from a first grader."

Roy wasn't just being a harsh, competitive parent. Alex had genuine difficulties with drawing, both in terms of having fine motor control over the use of pencils or crayons and in being able to draw meaningfully. When he'd use a pencil, even to write, he'd not press down hard enough on the paper. Over the years, Alex's occupational therapist spent hours teaching him how to draw a body. The most Alex could manage were stick figures, even though the therapist explained that the neck was not a stick, it had some breadth, or that hands were not just lines that ended in points, they were made of palms and fingers. "His drawings were so primitive," said Susan. And even today, many years later, ask Alex to draw a human shape, and the hands will most likely end in a circle with five smaller circles for fingers. "He still hasn't got it," said Roy.

David Cohen knows only too well of this inability of some autistic children to represent the human body. When I met Cohen, the head of child and adolescent psychiatry at the Pitié-Salpêtrière Hospital in Paris, in the autumn of 2011, he was still smarting from the ill effects of an article that appeared in the British journal *Lancet* in 2007, investigating the French practice of *le packing*, or packing therapy. The practice involves wrapping a child from feet to neck in cold, wet sheets, leaving only the head free to move. The child is then covered in a rescue cover and a dry blanket, thus allowing the body to warm up. Each such session lasts about an hour, and the therapy may involve multiple sessions over days or weeks. *Le packing* has been used in France as an additional therapy to help calm down severely autistic children with self-harming behaviors. *Lancet* illustrated the article with a photograph of the Chapelle de la Salpêtrière, a famous landmark in Paris at the entrance to the hospital, linking packing therapy with Cohen, though the story was hardly about his hospital unit (in fact, Cohen had yet to publish a single academic paper on packing therapy). Those who found the practice cruel and barbaric directed their ire at Cohen. Further negative publicity came in the form of a stern letter to the editor of the *Journal of the American Academy of Child and Adolescent Psychiatry*, in which a number of prominent autism researchers decried the "alleged therapy" as unethical.

But Cohen stands by the therapy. He pointed out to me, as he has done in articles he has published since the outcry in the *Lancet*, that the child is treated under the supervision of a *psychomotricien* (a specialist trained in psychomotor disturbances) and at least two members of the team caring for the child.

One of Cohen's patients, John, was an adolescent diagnosed with pervasive developmental disorder. He was catatonic when he was admitted to the hospital. Given the seriousness of his condition, electroconvulsive therapy (ECT) was an option, but John's parents refused it. Instead, they chose packing therapy, along with drugs (benzodiazepine and Prozac). John's condition improved while on this combination of therapies: he even agreed to draw after each session of packing therapy. The drawings showed something very interesting. After the second session, John wrote letters and words. It was only after session twelve that the first hints of a body appeared in the drawing: John drew a hand. He drew a stick figure after session sixteen, and a more realistic body after session twenty-three. It was as if the packing sessions were bringing John closer to his body.

Cohen views John's catatonia and indeed some of the sensorymotor disturbances observed in autism as a consequence of the inability of the brain to properly integrate all the various senses. The basic idea here is that the brain combines all the various sensations, both internal and external, such as touch, vision, vestibular and proprioception, to create a *body percept* (the sense of the body as an entity in itself—the bodily self), which then becomes the foundation for learning and behavior. Any disturbance of this multisensory integration process doesn't just disrupt perception of stimuli and one's own body, it also has behavioral and cognitive consequences.

Packing therapy, then, is helping reintegrate the senses, serving to "combine the body and the image of the body" and "to reinforce children's consciousness of their own body limits." In the parlance of the self, packing therapy is helping the self-as-subject form clear perceptions of the bodily self, the basic constituent of the self-as-object.

I asked Susan if this made sense to her, given her experience with Alex. She thought it did. "I believe that Alex's difficulties in understanding his body in relation to his physical environment caused delays in his language development, his abilities to self-organize, and caused his creative/imaginative and social impairments. These challenges must have had a big impact on the development of his sense of self," said Susan. "The reason his drawing of the human body was so primitive is because he may not have experienced fully the sensation of his limbs."

The consequences of sensory integration problems, in this way of thinking, could also cause a deficient theory of mind in children with autism. "Yes, they have that problem. Wouldn't you have it if you couldn't feel your body?" cognitive psychologist and computational neuroscientist Elizabeth Torres of Rutgers University asked me rhetorically.

● ● ○

Torres displays impatience with the status quo when talking about autism. To her, the practice of diagnosing autism by observing and cataloging behavioral aberrations "based on clinical observations with shifting criteria" is not helping. In fact, it's missing the point. According to Torres, the behavioral disturbances observed in those with autism are the result of one not being grounded in a stable perception of one's own body. Simple as that.

"People speak of behaviors in a very disembodied manner," Torres told me. "They talk about it as if it was something esoteric, but behavior is a combination of movements that flow continuously like a stream. Movements [in turn] are a combination of things that we do with a purpose and things we don't even know we are doing."

So, can one see anything in the way children with autism move that hints at an underlying problem? Torres's answer is an adamant yes.

Her assertion is based on measurements of movements of children diagnosed with autism spectrum disorder. Torres was looking for what she calls micromovements: barely perceptible fluctuations in how we move. For instance, if you were to reach out to touch a target on a computer touchscreen, at some moment *t* after you began moving your hand, it would reach a peak velocity *v*, and then it would slow down and eventually stop at the screen. Both *v* and *t* are examples of parameters that describe the motion of your hand. Interestingly, these parameters have tiny variations, or what Torres calls micromovements. If you reach out for the computer screen a hundred times, the values for *v* and *t* will be ever so slightly different each time. These variations in the parameters that describe our movements and the rates at which their statistical signatures change from moment to moment are unique to each person. Torres has argued that this variability in micromovements is a type of sensory input from the periphery of the body to the central nervous system.

It's a concept that goes back to work done in 1950 by Erich von Holst and Horst Mittelstaedt. We saw in the context of schizophrenia that the brain makes copies of motor commands, predicts the sensory consequences of these commands, and compares them to the actual sensations to generate a sense of agency. So, the brain would have to rely on some sort of error feedback from the body. Where exactly is the feedback coming from? "There could be error feedback from kinesthetic receptors in joints, tendons, and sensory muscle spindles.

Inputs from these receptors could indicate whether the moment-to-moment position of the arm corresponds to the intended position indicated by motor commands (efference copy). . . . Its use requires a learned association between motor commands and kinesthetic inputs."

The brain uses such kinesthetic signals, the argument goes, to build and maintain a stable percept of the body or an internal model of the body, so that it can then effectively issue motor commands and also accurately predict the consequences of carrying out the commands. This means that the micromovements should have valuable information, or a high signal-to-noise ratio.

When Torres studied variations in these micromovements in children of varying abilities and ages, something very intriguing emerged.

The signal-to-noise ratio increased with age. In three-to-four-year-old typically developing children, the kinesthetic error feedback signals were very noisy. But in four-to-five-year-olds, the signals were significantly less noisy. In adults, these kinesthetic inputs were reliable and predictable. But autism affects this progression. The feedback from the micromovements is extremely noisy in both children and adults diagnosed with autism. If the brain is indeed working with internal models of the body, then these inputs are not helping to keep these models updated. They provide very little information for the brain about prior behavior on which it can base future behavior. It's as if the child with autism is struggling to make sense of every experience. "That's the way they must be experiencing the world. It's constantly new. They can't make sense of it. They can't come to a stable percept that enables one to predict ahead," Torres said.

The self-as-subject is unable to get a fix on the self-as-object.

The body anchors the self-as-object; it's our point of reference.

Everything we perceive is in relation to it. According to Torres, a glitch during development can disturb this point of reference, and depending on when that glitch happens, it can lead to a bewildering array of consequences as the child grows up—which helps explain the huge range of behavioral symptoms that get clumped under autism spectrum disorder, ranging from problems processing sensations to theory-of-mind disturbances and difficulties in relating socially to others. "If you don't have that point of reference, and everything is always new to you, there is no anchor," said Torres. "This must be happening to people with autism, because their information about the body is noisy and it's random. We have measured very precisely; it doesn't depend on my opinion, and it is what it is. And it really is present in every autistic person we have seen and it gets worse with age."

Fortunately, said Torres, this also means that detecting the noisy system early, with objective measurements, not a clinician's subjective observations—as well as developing therapies to train the body and reduce the noise—is likely to be extremely beneficial.

● ● ○

Torres's work sits well with the idea of the Bayesian brain, the idea that the brain could be making probabilistic inferences about the likely causes of sensory inputs. We saw how this framework could be applied to emotions to explain depersonalization disorder. Applying it to autism is proving valuable too.

If you accept that the body's imperative is to survive, then the brain's job (in close collusion with the body) is to maintain the body in a state suitable for survival. For any given biological organism, survival means existing in one of a finite set of physiological states. For instance, if you take internal parameters like blood pressure and heart rate, and external parameters like temperature, and define body states based on these parameters, then there are only a limited set of states in which all these parameters are within acceptable limits. Looked at another way, "there is a high probability that a [biological] system will be in any of a small number of states, and a low probability that it will be in the remaining states," according to Karl Friston of University College London. The process of staying within the bounds of physiologically viable states, again, is known as homeostasis.

Friston posits that the brain achieves homeostasis by minimizing what he calls free energy, which allows biological systems (or indeed any system that can learn and adapt) to "resist a natural tendency to[ward] disorder." In the context of a population out in the wild, systems that minimize free energy survive; those that don't, die.

Friston has shown that minimizing free energy for a biological system is equivalent to minimizing the amount of surprise it encounters as it navigates its environment. "Biological agents must avoid surprises to ensure that their states remain within physiological bounds," he says.

To review, the way a Bayesian brain avoids surprises is by maintaining an internal model of the body, the environment, and of itself. This is a probabilistic model that can generate a number of predictions about the causes of sensory inputs based on a set of prior beliefs as to the causes. Then, given actual sensory data, the brain assigns new probabilities to its predictions, and the prediction with the highest probability is what we perceive as the cause of those sensations. Of course, now the brain has a new set of prior beliefs, updated to reflect its new understanding of the body, brain, and environment. Where does surprise enter this picture? Well, if you found yourself in a state that was extremely surprising and detrimental to your existence (an example from Friston is that of a fish out of water, an extremely surprising state for the fish), your brain would initiate action to suppress that element of surprise—the action could either involve modifying its internal models or making the body move (a fish out of water had better flip-flop its way back into water; modifying internal models will not help). So, the greater the discrepancy between the brain's predictions about the causes of sensory inputs and the actual sensory inputs, the greater is the element of surprise. Minimizing surprise is akin to minimizing prediction error—which implies a brain whose models are in tune with external and internal reality.

According to Elizabeth Torres, in autism the brain's ability to modify its prior beliefs based on the actual sensory input is impaired, because of the high levels of noise and low levels of useful information in the error feedback signals. So, people with autism may live with a constant element of surprise.

Being prone to continual surprise, an autistic person's world must appear magical—but not in a good way, argue Pawan Sinha of MIT and his colleagues. Magicians, they point out, rely on being able to surprise their audience. When people are unable to predict a magician's next move, it leads to wonder and astonishment. But in the real world, if you are unable to predict the causes of things, it can be debilitating. "A magical world suggests lack of control and impairs one's ability to take preparatory actions." Their hypothesis is that an impaired predictive brain might be responsible for the wide range of seemingly disparate symptoms that characterize the autism spectrum disorder, which echoes Torres's ideas.

Take an autistic child's insistence on sameness. Being unsure of your environment is a recipe for anxiety even in neurotypicals, and this might be amplified in autism.

Susan provided some additional perspective, having experienced this with her son, Alex, as well as with other kids in his school who have autism. "Change is especially scary for folks on the spectrum. Kids on the spectrum thrive on routine and repetition. Reading the same book over and over again, watching the same movie multiple times, having a narrow repertoire of preferred foods, going to a small number of restaurants, opting for the same item from the menu *every* time, et cetera. They thrive on predictable schedules and firm expectations."

If an autistic person is constantly confronting an unpredictable environment, then their desire not to veer too far from certain behaviors may be due to their need to increase predictability and reduce anxiety.

Even hypersensitivity to light, sounds, and other stimuli could be explained: impaired mechanisms for the prediction of causes of environmental stimuli, or imperfect updating of prior beliefs, could make stimuli seem endlessly novel.

It's not just problems with sensory perception that can be explained using this framework. Sinha and his team also hypothesize that an impaired predictive brain could lead to problems with theory of mind. Reading someone else's mind is akin to predicting the causes for their observed behavior (that is, their intentions and desires), while keeping in mind prior knowledge about the person. "Theory of mind is inherently a prediction task," they wrote. In science-speak, an impaired predictive brain is a parsimonious explanation for autism.

From the perspective of the self, some neuroscientists and philosophers argue that the predictive brain could explain even those aspects of the self that are considered basic, prereflective, and prenarrative.

"The brain has to model everything it encounters, including itself.

It's a straightforward consequence of thinking of the brain as this statistical machine. In that sense, the self is just a representation maintained in the brain, just like any other representation that we have of objects in the world," philosopher Jakob Hohwy of Monash University in Melbourne, Australia, told me.

For instance, as we saw in the chapter on schizophrenia, the sense of agency is the product of a predictive brain. We also saw that the vivid connection to our emotions can be lost due to prediction gone wrong, leading to depersonalization, making us strangers to ourselves. And as we will see in the next chapter, even other properties that define the self—the sense of *mineness*, the feeling that I inhabit a body that feels like *mine*—could potentially be explained by thinking of the brain as an inference engine. It certainly takes the air out of the balloon for theories that privilege the self over, say, visual perception. "We have a deflation of the concept [of the self]," said Hohwy. "There is nothing special about the self; it's just another cause of our own sensory inputs."

So, anything that constitutes the self-as-object—aspects of the self that are experienced by the self-as-subject—could be thought of as perceptions arising out of the brain's predictive mechanisms. Whether you are a neurotypical or someone with autism, you may be your brain's best guess as to causes of all your internal and external sensory signals put together.

This gives me pause. My thoughts wander to Alex as a little boy. I have known him most of his young life. I wonder about Alex as a two-year-old, who insisted on sameness (whether in the food he was eating or the clothes he was wearing), and his penchant for lining up his toy cars, and how it rattled him to see the order disrupted. He shied away from people, not wanting to be held or picked up. Were these behaviors attempts at avoiding surprise, his way of creating a predictable environment? Alex has overcome some of his fears as he's entered adolescence. He lets himself be hugged, for example. Understanding why children with autism insist on sameness and predictability can only help social interactions. Maybe those with autism have an altered self-experience and have trouble reading other people's minds. But one could also argue that neurotypicals have trouble reading autistic minds. Communication, by definition, is a two-way street, even if at times it is between potentially different minds.

7

WHEN YOU ARE BESIDE YOURSELF

OUT-OF-BODY EXPERIENCES, DOPPELGÄNGERS, AND THE MINIMAL SELF

This proposition [that] . . . I am, I exist, is necessarily true each time that I pronounce it. . . . But I do not yet know clearly enough what I am.

—René Descartes

"Owning" your body, its sensations, and its various parts is fundamental to the feeling of *being someone.*

—Thomas Metzinger

My cousin's son, Ashwin, a youthful thirty-one-year-old, died of brain cancer recently. The first indication of a potential problem came in August 2009. Ashwin had had a major seizure. Neurosurgeons in New Delhi found and removed a benign tumor from the left temporoparietal region. Within months of the operation, he began having seizures again. Scans revealed nothing new, so he was put on anticonvulsant medications. Ashwin learned to recognize the onset of seizures, usually pins and needles in his right arm and leg. If he was driving, he'd pull over and take some deep breaths (his mother's instructions), and wait for the seizure to pass. Moments later it would be over. Then, in early 2013, he was driving to work when something very odd happened. He immediately stopped his car on the roadside and called his mother.

"Mom, I had a very strange experience," he told her. "I saw another Ashwin in front of me." He was in no doubt as to what he had seen and experienced: his own self facing him. He was even aware of the emotional state of this double. Ashwin told his mother that the second Ashwin was angry, resentful, frustrated (an emotional state that mirrored how he used to be in his twenties, my cousin told me). Thankfully, the double disappeared, and Ashwin could drive again. His neurologist attributed the experience to a seizure and adjusted his medication.

Within a year, however, Ashwin's condition worsened. His tumor returned, this time with a vengeance. It was malignant, and in the left frontotemporal region, its tentacles spreading into the left insular cortex. Surgery and radiation therapy bought him some time, but not much. Ashwin passed away very suddenly one evening.

What Ashwin experienced that morning in his car is a phenomenon called the doppelgänger effect. It's a complex hallucination that involves the feeling that there is another illusory body of oneself nearby, as happened in his case. While Ashwin remained in his physical body, often the person hallucinating can find that his or her center of awareness—the sense of being in a body looking out—can shift from the physical body to the illusory body. The person switches perspective, seeing the world either from the physical body or the illusory body, sometimes moving back and forth in rapid succession. Another distinguishing characteristic of the doppelgänger effect is often the presence of strong emotions. One of the most cited accounts in the medical literature of the doppelgänger experience is of a young man who jumped off a four-story building to reconcile his self with his body.

● ● ○

More than two decades ago, Peter Brugger, as a PhD student in neuropsychology at the University Hospital Zurich in Switzerland, was developing a reputation as someone interested in scientific explanations of so-called paranormal experiences. A fellow neurologist, who had been treating a twenty-one-year-old man for seizures, sent the patient to Brugger. The young man, who worked as a waiter and lived in the canton of Zurich, had very nearly killed himself one day, when he found himself face-to-face with his doppelgänger.

The incident happened when the young man had stopped taking some of his anticonvulsant medication. One morning, instead of going to work, he drank copious amounts of beer and stayed in bed. But it turned out to be a harrowing lie-in. He felt dizzy, stood up, turned around, and saw himself still lying in bed. He was aware that the person in bed was him, and was not willing to get up and would thus make himself late for work. Furious at the prone self, the man shouted at it, shook it, and even jumped on it, all to no avail. To complicate things further, his awareness of being in a body would shift from one body to the other. When he was inhabiting the supine body in bed, he'd see his duplicate bending over and shaking him. That's when fear and confusion took hold: Who was he? Was he the man standing up or the man lying in bed? Unable to take it, he jumped out the window.

When I visited Brugger in the autumn of 2011, he showed me a photograph of the building from which the man had jumped: he had been extremely lucky. He had leapt from a window on the fourth floor and landed on a large hazel bush, which had broken his fall. But he had not really wanted to commit suicide, said Brugger. He had jumped to "find a match between body and self." After getting treatment for his fall-related injuries, the young man underwent surgery to remove a tumor in his left temporal lobe, and both the seizures and the bizarre experiences stopped.

● ● ○

Doppelgängers are the stuff of literature: from Edgar Allan Poe's "William Wilson," in which William, tormented by his double, stabs him, only to realize that he himself is bleeding, to Guy de Maupassant's short story "Le Horla," in which the main character murders his double, but laments at the end, "No . . . no . . . of course not . . . of course he is not dead. . . . So then—it's me, it's me I have to kill!," fictional doubles abound.

Broadly, such hallucinations are classified as *autoscopic phenomena* (from "autoscopy"; in Greek, *autos* means "self" and *skopeo* means "looking at"). The simplest form of an autoscopic phenomenon involves feeling the presence of someone next to you without actually seeing a double—a sensed presence. Olaf Blanke, a neurologist at the Swiss Federal Institute of Technology in Lausanne, Switzerland, told me that a sensed presence is like experiencing a full-body phantom: if a phantom limb is the continued sensation of having a limb that has been amputated, then a sensed presence of a body is its full-body analogue.

T. S. Eliot immortalized such an extracorporeal presence in his poem *The Waste Land*: "Who is the third who walks always beside you? / When I count, there are only you and I together."

As it turns out, Eliot was inspired by accounts of the Antarctic explorer Ernest Shackleton, who wrote in his diaries that he and expedition team members Frank Worsley and Tom Crean, on the last leg of an unimaginably dangerous and difficult journey to find help to save the other stranded members of their trans-Antarctic expedition, began feeling the presence of a fourth person. Shackleton wrote, "I know that during that long and racking march of thirty-six hours over the unnamed mountains and glaciers of South Georgia it seemed to me often that we were four, not three. I said nothing to my companions on the point, but afterwards Worsley said to me, 'Boss, I had a curious feeling on the march that there was another person with us.' Crean confessed to the same idea. One feels 'the dearth of human words, the roughness of mortal speech' in trying to describe things intangible, but a record of our journeys would be incomplete without a reference to a subject very near to our hearts." We now know that it's not uncommon for oxygen-deprived mountaineers to report sensing the presence of another.

Autoscopic phenomena can go beyond just a sensed presence. There is the doppelgänger effect, in which a person may hallucinate that they are actually seeing another "me"—a visual double. Often, the hallucination is very emotional, and the person's sense of location and identity switches between the real and the illusory bodies, as experienced by Brugger's twenty-one-year-old patient.

Probably the most widely experienced and best-known form of autoscopic phenomena is the out-of-body experience (OBE). During a classic full-blown OBE, people report leaving their physical body and seeing it from an outside perspective, say from the ceiling looking down at the body lying in bed.

During my discussions with Michaele about her husband Allan's battle with Alzheimer's disease, I mentioned to her that I was also writing about out-of-body experiences. As it happened, well before she met Allan, Michaele had an intense out-of-body experience. She was in her thirties and pregnant with her fourth child. When it came time to have the baby, a son, she chose a home birth, in the presence of a midwife and a physician. Her water broke one night, and the next morning her physician went over to the local abortion clinic to get a tablet of Pitocin, which can be used to induce labor. Michaele put the pill under her tongue and soon went into labor. She had chosen not to take painkillers. At the very peak of the process, just as she had pushed her baby out, the pain became unbearable. Michaele felt herself leave her body. "I literally was up at the corner of the ceiling, looking down at the whole scene, watching everything happen," she told me. "I just left my body. It got so intense that I went above, and as soon as it was over, I was back, right back in my body again. It was the weirdest thing." She thinks the whole episode may have lasted just a few seconds, but more than three decades later, the experience is still etched in her mind. "It's not something I have talked about a lot," she said. "I have only told a few people that I feel would understand."

Many people who have such experiences are reluctant to talk about it. OBEs give the person a strong sense of dualism of body and mind: your center of awareness, which is usually anchored in your body, seems to float free of it. We saw earlier how the bodily self is the foundation for our sense of self, and disruptions of the bodily self can cause BIID, schizophrenia, and perhaps even autism. In all these cases, however, the center of awareness remains anchored to the body, however impaired the perception of it may be. OBEs mess with this center of awareness—suggesting a Cartesian duality. But if you examine OBEs closely, it turns out that the duality is an illusion, a product of a brain that fails to correctly integrate all the signals from the body. Despite their vividness, OBEs are hallucinations caused by malfunctions in brain mechanisms; elucidating these mechanisms gets us closer to understanding how the brain constructs the self.

● ● ○

Back at the University Hospital Zurich, Peter Brugger tried gamely to induce in me an out-of-body illusion. We were wandering the corridors of the hospital. I was wearing virtual-reality goggles. Brugger was walking about three feet behind me, filming me using my notebook computer's webcam and feeding the video into the goggles I was wearing. So, instead of seeing where I was going, I was seeing myself from behind, walking about three feet in front of me. We must have been a sight as we walked past curious interns and hospital staff. Brugger, looking like an absentminded professor with his white lab coat and wild, graying hair, holding aloft an open notebook computer, and me walking in front, blind but for what I was seeing in the VR goggles.

The setup didn't quite work. We should have been using a good video camera, which we didn't have at the time, and longer wires so that Brugger could have been farther behind me. But I did feel weird walking around watching myself from behind.

In 1998, when Brugger first tried the experiment, he wore such goggles for an entire day, and had someone walk about twelve feet behind him, filming him with a video camera. So, if Brugger was picking a flower, or putting a letter in a mailbox, he'd see himself doing the act from an outside perspective. "This was extremely strange. I lost the sense of where I actually was," he told me. "I was where I saw the action, rather than where I was actually executing the action." Brugger was having an out-of-body illusion: the sense of where he was located had shifted several feet, from being in his physical body to being in the virtual body.

But Brugger never actually performed the experiment in a rigorous laboratory setting and so never published the results, though it did get mentioned in an article in *Science*.

He credits American psychologist George Malcolm Stratton (1865–1957) as the inspiration for this experiment. Stratton had spent a good part of his career at the University of California, Berkeley. He is best known for "perhaps the most famous experiment in the whole of experimental psychology." Stratton fashioned a contraption that allowed him to see upside down. He walked around with this device on his right eye. He blacked out the left eye, because seeing upside down with both eyes was extremely disorienting. For three days and a total of 21.5 hours, he did nothing but use this device. When he went to bed, he strapped his eyes shut. While the primary motivation for the experiment was to understand visual perception, Stratton experienced other subtle changes in bodily perception. For instance, if he stretched out his hand to touch something, because he was seeing everything upside down, the hand would enter the visual field from above rather than below. Soon, "parts of my body . . . were seen to be in another position."

Stratton realized that he was onto something. In 1899, he published another paper, in which he described a crazier experiment, this time with mirrors. He built a frame that he affixed to his waist and shoulders. The frame held a mirror horizontally above his head. He used the frame to position another mirror at a forty-five-degree angle in front of his eyes, so that it reflected the image from the overhead horizontal mirror directly into his eyes. The net effect was that Stratton was seeing himself and the space around him from the perspective of someone looking down at his head from above. He made sure that no other light entered his eyes. Again, he walked around with this contraption for three days, for a total of twenty-four hours, with his eyes blindfolded when he was not experimenting and when he slept. Doing so, he was able to create a disharmony between sight and touch: when he'd reach out to touch something, his hands felt the touch, but his eyes told him that the touch was somewhere else entirely. It was now up to the brain to bring everything back into harmony, with interesting consequences.

Because Stratton was seeing his own body from above and nothing else, he had to pay close attention to this visual image to guide his actions and movements. By afternoon on the second day he began to notice that the reflected image sometimes felt like his body. This feeling became more persistent on the third day, especially when he was walking with ease and speed, not making any special effort to differentiate between where he was perceiving his body to be and where he "knew" his body to be. "In the more languidly receptive attitude during my walk, I had the feeling that I was mentally outside my own body," he wrote. Stratton had induced in himself an out-of-body experience.

• • ○

Out-of-body experiences, autoscopic hallucinations, and doppelgänger phenomena are probably our best window on some very basic aspects of our sense of the bodily self. It's become increasingly clear that the brain's representation of one's body and our conscious experience of it underpin self-consciousness. Having a bodily self, or being embodied, means several things. At the very fundamental level, it situates the center of our awareness. You are in a body that feels like it is yours—this is the sense of self-identification and body ownership. You also feel that the body occupies a certain volume in physical space and you are located in that volume—the sense of self-location. Finally, you look out at the world from a point behind your eyes and you have the sense that this vantage point is yours and yours alone—you have what philosophers call a first-person perspective on the world.

The rubber-hand illusion is a classic example of how aspects of this bodily self can be disrupted. As we saw in chapter 3, when an experimenter strokes a visible rubber hand and the hidden real hand synchronously, the rubber hand is temporarily incorporated into one's bodily self. We feel touch at the location of the rubber hand and there's a sense of ownership of this otherwise lifeless object.

Henrik Ehrsson's team at the Karolinska Institute in Stockholm, Sweden, got people to experience the rubber-hand illusion while they lay inside an fMRI scanner. The findings were revealing. The strength of the illusion was strongly correlated with activity in the premotor cortex, a region in the brain that forms a network with the cerebellum and with parietal areas that process vision and touch. Parts of the parietal brain regions integrate vision, touch, and proprioception, and it's well known that people with parietal lesions sometimes deny ownership of their limbs.

Neuroscientists think that the so-called multisensory integration of various sensations is responsible for giving us a sense of ownership over our body and body parts. Normally, vision, touch, and proprioceptive sensations all match up. They are congruent, and it's this congruency that's key to giving a body part a sense of *mineness*. During the rubber-hand illusion, proprioceptive distortions are kept to a minimum by keeping the real hand relaxed and not too far from the rubber hand. The brain erroneously integrates the misleading visual sensations and the real sensations of touch, and decides that the rubber hand is real. That's why we can lose ownership of the actual hand and gain ownership of the rubber hand. This switch in ownership has real physiological consequences: for instance, the temperature in the real hand drops by nearly 1 degree Celsius (about 2 degrees Fahrenheit)—an autonomic nervous system response that's not under conscious control.

In Ehrsson's lab, I got to experience the rubber-hand illusion for the first time (having failed in earlier attempts). Arvid Guterstam, a postdoc in Ehrsson's lab, played host and subjected me to the illusion. Having done it innumerable times, he was quite the expert. I felt the illusion of owning the rubber hand rather strongly. But then Guterstam did something that jolted me further. Once I began feeling the touch in the location of the rubber hand, he lifted his brush a couple of inches above the rubber hand and continued moving the brush synchronously with the movement of the brush on my real hand.

"What was that?" I said. "What's happening? This is really weird."

He was moving the brush in the air and I was feeling the touch of the brush in the space above the rubber hand.

It turns out the neurons in the premotor cortex have what's called a receptive field—they fire not just when a body part is touched but when the proximal space around that body part is touched (this is called peripersonal space). My brain had remapped the location of my hand and centered it on the rubber hand. The space above the rubber hand had become my peripersonal space, and consequently a brush-stroke in the space above the rubber hand was now registering as a touch at that location.

Ehrsson's team has also shown that you don't even need a rubber hand to experience the illusion: just the brush strokes on the hidden real hand combined with synchronous movements of a brush in empty space, in a manner that's suggestive of a hand there, is enough to generate the illusion of being touched at a location where there is no real hand.

Scientific explanations aside, I was thrilled to have finally experienced the illusion, and said so.

"You seem to have an easily fooled brain," Guterstam quipped.

Fooling the brain to take ownership of a rubber hand is just one piece of the puzzle that is bodily self-consciousness. A hand is just a constituent of the bodily self. How much more can one manipulate the bodily self? Much, much more, as it turns out.

● ● ○

As a young man in the late 1970s and early 1980s, Thomas Metzinger felt conflicted about telling anyone about his out-of-body experiences. One of those happened when he was studying to become a philosopher, and intensely curious about altered states of consciousness. He was attending a highly regimented meditation retreat in the Westerwald, about sixty miles northwest of Frankfurt, Germany. Ten consecutive weeks were filled with yoga, breathing exercises, and individual and group meditation sessions. Metzinger immersed himself in all that was asked of him. One Thursday, the retreat organizers had baked a cake—to celebrate the teacher's birthday. It was a rich, greasy cake. Metzinger ate some of it. Feeling unwell, he went to bed and fell asleep.

He woke up wanting to scratch his back, and realized he couldn't move. His body was paralyzed. It was then that he felt himself spiraling out of his own body, up and in front of his bed. It was dark, so he didn't actually turn around and see his body lying in bed. He was scared, but something scarier was to follow.

He suddenly realized that there was someone else breathing heavily in the room. "And then I panicked," Metzinger told me as we sat at the dining table in his home in rural Germany, a few tens of miles east of Frankfurt. "Somebody was there; I couldn't move my body, I was dissociated from my body. It was very unpleasant." There was, of course, no one else in that room, and only many years later would Metzinger find explanations for such experiences in the scientific literature. It turns out that in certain dissociative states, you cannot recognize self-generated sounds as self-generated; in Metzinger's case, he lost a sense of ownership of the sounds of his own breathing, hence the hallucination of someone breathing near him.

Metzinger alerted his meditation teachers, but to his dismay all they did was put him under a cold shower and tell him to meditate less (today, as someone who advocates meditation training in schools, Metzinger is concerned and critical that many meditation centers do not have staff trained to deal with altered states of consciousness or psychiatric emergencies).

Soon afterward, Metzinger moved to a remote region south of Limburg, to concentrate on writing his doctoral dissertation on the mind-body problem and also to deliberately confront himself with the consequences of solitude and boredom—a personal project. As a poor student, with little money to even call his friends in Frankfurt, he lived alone in a 350-year-old house, taking care of sheep and nineteen fishponds. He meditated a lot. And he had a few more unexpected, spontaneous out-of-body experiences. But by now, his curiosity and analytical mind had taken over: he wanted to understand his experiences. His extensive study of the scientific and philosophical literature was showing a complete lack of evidence that consciousness could be dissociated from the brain. Yet there he was, having extremely vivid experiences of apparent dualism in which his conscious self was seemingly separated from his own body. And he knew he could tell no one except his closest friends.

So, as a budding philosopher of mind and cognitive science getting grounded in empirical data, he tried conducting his own experiments while in those altered states, to see if brain and consciousness could indeed be separated and whether that would lead to conclusive, verifiable observations. He learned to control his initial fear during his OBEs, but not entirely. Despite his efforts, he uncovered not a shred of evidence that his conscious self had actually dissociated from his body.

Meanwhile, he had conversations and exchanges with other researchers. One British psychologist, Susan Blackmore, after fierce and extended discussions, finally managed to convince him that his OBEs were actually hallucinations. She quizzed him about how he moved from his physical body, which was lying in bed, to the windowsill during an out-of-body experience. Did he walk over there? Did he fly? Metzinger realized that his movements were unlike anything that happens in real life. "Sometimes, it's almost as if the moment you think you want to go there you are already there," he told me. Blackmore argued that he was hallucinating, moving between mental representations of, say, the bed and the window, jumping or gliding from landmark to landmark in his mind. Metzinger realized that he was not moving in his bedroom but within an internal model of his bedroom created by his brain.

Another really strange experience convinced him even more that he was indeed hallucinating. He had an OBE, and when he returned to his body he ran to wake up his sister, to tell her of his experience. "It's quarter to three, can't it wait until breakfast?" she told him. But then an alarm went off, and Metzinger woke up—again. He wasn't in Frankfurt in his parents' house with his sister. Rather, he had been taking an afternoon nap in a house that he shared with four other students. He had experienced what dream researchers call "false awakening": a dream that he had woken up. But prior to the false awakening, he had dreamed he had an OBE. "It began to dawn on me that there are multiple transitions between different classes of altered states of consciousness," he said. He had been having such vivid out-of-body experiences that he had even begun dreaming about them.

Metzinger's OBEs stopped after six or seven such episodes. But they have informed his thinking about how the brain might be causing them and what it tells us about the self, eventually resulting in his definitive monograph: *Being No One: The Self-Model Theory of Subjectivity.* The work caught the attention of Olaf Blanke, the neurologist whom I met at the Swiss Federal Institute of Technology in Lausanne.

● ● ○

In 2002, Blanke had induced repeated out-of-body experiences in a forty-three-year-old woman. He had been treating her for drug-resistant temporal-lobe epilepsy. Brain scans did not show any lesions, so Blanke resorted to surgery to figure out the focus of her epilepsy. His team inserted electrodes inside the cranium to record electrical activity from the cortical surface directly, rather than from outside the skull as you would if you were using standard EEG. During this procedure, the woman volunteered to have her brain stimulated using the implanted electrodes. This technique allows surgeons to double-check that they've really found the cause of the seizure, while also ensuring that they don't excise some key brain region. And not just that. The procedure, pioneered by Wilder Penfield, is often the best way to find out the function of different brain regions, and much of what we have learned about the brain has come from courageous patients who have let themselves be stimulated while conscious. It was during such a procedure that Blanke found that there was one electrode, placed on the right angular gyrus, that, when stimulated, caused the woman to report some rather weird sensations.

When the stimulating current was low, she reported "sinking into the bed" or "falling from a height"; when Blanke's team increased the amperage, she had an out-of-body experience: "I see myself lying in bed, from above," she said. The angular gyrus lies near the vestibular cortex (which receives inputs from the vestibular system that's responsible for our posture and sense of balance). Blanke concluded that the electrical stimulation was somehow disrupting the integration of various sensations such as touch with vestibular signals, leading to the woman's OBE.

The next step in studying OBEs in a controlled setting was to try to induce full-body versions of the rubber-hand illusion in healthy subjects in a laboratory. In 2005, Metzinger proposed an experiment to do just that. He teamed up with Blanke and Blanke's then student Bigna Lenggenhager. The setup they used was simple and elegant. A camera filmed a subject from behind, and the images were sent to a 3-D head-mounted display that the subject was wearing. The subject could see only what was being shown in the display, which was the back of his or her own body, seen in 3-D and about seven feet in front (this was analogous to my seeing the rubber hand rather than the real hand). The experimenter would then stroke the person's back with a stick. The subjects would feel the stroking on their backs, but would also see themselves being stroked in the head-mounted display. The stroking was either synchronous or asynchronous (to make it asynchronous, the video feed was delayed a smidgen, so the subject felt the touch first but saw the virtual body being stroked an instant later). Again, this is not unlike the rubber-hand illusion experiment—nor were the results dissimilar. In the synchronous condition, once the illusion set in, some subjects (but not all) reported feeling the touch in the location of the virtual body about seven feet in front of them and that the virtual body felt like their own.

A few years later, Blanke's team upped the ante. They rigged a setup that allowed them to conduct the same experiment inside a scanner. The subject was lying down, and a robotic arm stroked the subject's back. Meanwhile, the subject viewed through a head-mounted display a video of a person being stroked on the back. The robotic arm's stroking was either synchronous or asynchronous with stroking of the virtual person seen on the display. Again, in some subjects, their sense of location and sense of body ownership were shaken up. One of the most striking outcomes was when a subject reported "looking at my own body from above," even though the subject was lying prone, face-up, in the scanner.

"That was for us really exciting, because it gets really close to the classical out-of-body experience of looking down at your own body," said Lenggenhager, who is now working in Peter Brugger's group at the University Hospital Zurich.

The subjects were scanned during their experiences, and the scans revealed that their sense of being out-of-body was correlated with activity in the temporoparietal junction (TPJ), a site that integrates touch, vision, proprioception, and vestibular signals. Here was some objective evidence that self-location—where you perceive yourself to be—has to do with neural activity in the TPJ.

When I visited Lausanne, Blanke's student Petr Macku offered to try the illusion out on me and I gladly accepted, for that was partly why I was visiting. He used the same equipment, except for a scanner—but I must have been too tense (having just arrived from Paris), and possibly expecting too much, because the illusion didn't work on me. The other likely explanation is that the full-body illusion is a weak effect, and does not work on everyone. I did feel a bit strange, but that was it.

I was subjected to yet another full-body illusion in Henrik Ehrsson's lab in Stockholm (where I had successfully experienced the rubber-hand illusion). In this case, I stood facing a life-size mannequin, mirroring its outstretched hands. The mannequin had cameras for eyes and it was gazing down at its abdomen and hands. The camera output was fed into a head-mounted display that I was wearing. So, I was seeing the mannequin's abdomen and hands. Arvid Guterstam, the expert manipulator of rubber hands, again did the honors: using two big paintbrushes, he stroked my abdomen and hands, while doing the same to the mannequin's abdomen and hands, synchronously. I was feeling the touch on my body but seeing the mannequin's body being touched. Nothing much happened when he was stroking the abdomen (so much for my easily fooled brain), but after a couple of minutes, when he would brush my fingers, I would feel as if the mannequin's fingers were being touched. I was identifying with the mannequin's fingers as my own, if not its full body.

Henrik Ehrsson's team carried out a similar experiment inside a scanner, and subjects reported identifying with the mannequin's body. Many said they felt that the mannequin's body was their own. The scans showed that activity in the ventral premotor cortex in both hemispheres, along with activity in the left intraparietal cortex and the left putamen, was correlated with feelings of body ownership, with the correlation being strongest for the ventral premotor cortex. It's known from studies of macaque monkeys that neurons in these regions also integrate vision, touch, and the proprioceptive sense.

What's clear from these studies is that aspects of our sense of self that we take as given and immutable—a sense of body ownership, a sense of where the self is located, and even the perspective from which the self observes—can be disrupted, even in healthy people.

It's also becoming evident that self-location, self-identification, and first-person perspective are the result of different brain regions integrating the various sensations—touch, vision, proprioception, and vestibular sensations—to construct these aspects of selfhood. For instance, in the Ehrsson lab version of the full-body illusion, they were able to manipulate a sense of body ownership, and identify the correlated brain regions (mainly the ventral premotor cortex). The Blanke lab version of the illusion messed with perspective and self-location, and that potentially explains why they found a different brain region—the TPJ—as the main culprit.

The exact brain regions aside, the strong message here is that these attributes of self-location, self-identification, and first-person perspective are constructed by the brain. The brain creates a body-centered frame of reference, and everything we perceive is then intimated to us in terms of this frame of reference.

So far, we have been talking of the integration of various external sensations with sensations that tell the brain about the orientation of the body and the location of body parts. But there is another important source of sensations—something we are normally unaware of—which are signals from inside the body, especially the viscera (which contain information about the beating heart, blood pressure, and the state of the gut, for example). We saw in an earlier chapter how these internal sensations are key to emotions and feelings, and that malfunctions in this pathway can lead to depersonalization and feelings of being estranged from oneself. It turns out that in order to anchor the self to the body, the brain has to integrate signals from within the body with external sensations, and with sensations of position and balance. When something goes wrong with brain regions that integrate all these signals, the results are even more dramatic than out-of-body experiences. They lead to the doppelgänger effect, the kind Ashwin experienced sitting in his car and that caused Brugger's patient to jump out of a fourth-floor window in Zurich.

One of the most striking aspects of the doppelgänger effect is the presence of strong emotions—and what this reveals about the brain mechanisms involved. Of all the accounts that I've heard or read about, none had a stronger emotional content than Chris's experience, in which his double communicated with his brother, who had just died of HIV/AIDS.

● ● ○

Chris grew up in the San Francisco Bay Area. He was seven years older than his brother, David. As children, Chris and David fought all the time, "as brothers often do." It wasn't until Chris moved out of their parents' house that the brothers realized they missed each other. Over the next decade their relationship deepened. They also had a natural comedic chemistry; they were the "Martin and Lewis of the family," with David being Jerry Lewis to his elder brother Chris's Dean Martin. The gags were constant. They made outrageous bets with each other. David took on a bet, for instance, that he could eat an entire two-pound block of cheddar cheese all at once—an effort that had the family in stitches around the kitchen table as they watched David trying to stuff the cheese into his mouth, eventually just laughing hysterically and drooling melted cheese.

They played a relentless game of "gotcha." Chris recalled one incident where he got David well and good. David at the time was sporting an Afro hairdo and was sitting and watching TV with the family. Chris had been working on the water heater outside the house when he spied a large alligator lizard, a native California species. Chris caught the lizard and put it into a pocket of the overalls he was wearing. He came back into the house and discreetly maneuvered himself behind David and dropped the lizard on his Afro.

David knew Chris was up to something but was blasé about it. "Then the lizard took off. It ran right across the top of his head, down his face, and jumped onto his chest. My brother just screamed, and levitated off of the chair," Chris told me. "I swear he was two feet off the ground, screaming all the way across the room." Once David realized he'd been had, he laughed too, and then everyone spent the next forty-five minutes looking for the lizard. They never found it, dead or alive.

When David turned sixteen, he asked if he could come and spend the weekend with Chris. It was uncharacteristic of him to come for a whole weekend, so Chris knew something was up, and even had an inkling of what. Toward the end of the visit, a nervous David said, "Chris, I have to tell you something." Chris said OK, tell me.

"I'm gay," said David.

"Tell me something I don't know," said Chris.

"What? You know?"

"I've known since you were nine. Come on, how could I not know? I'm your brother," said Chris.

Eventually, David came out to his parents, who were crushed, especially their mother. Chris got mad at his parents, and confronted them about whether they saw any difference between him (the straight son) and David. "It kind of smacked them a little bit, stung a bit," Chris told me. But soon enough, the family came together.

A few years later, David told Chris that he had contracted HIV. "He was running around with a wild crowd over in San Francisco," said Chris. "There was a lot going on of what you might traditionally expect of San Francisco in the late '70s and early '80s." This was the early days of the HIV/AIDS epidemic, and HIV drugs weren't as effective. David knew he was dying, so he asked Chris to write his eulogy for the impending funeral.

"You can't die, I'll be alone," Chris told him. "There won't be Lewis to my Martin." Even decades later, as he told me this, Chris's voice broke; he could not contain his sadness.

David died with family by his bedside holding him. Chris and their father spoke at the funeral, with their father speaking of David's serious side, while Chris narrated the Martin-and-Lewis stories. And in accordance with David's wishes that they play "Amazing Grace" at the funeral, a Scottish piper in a kilt played the tune until the service was over.

About two months later, Chris woke up from his sleep. It was early in the morning. He got off the bed, stood up, and walked toward the end of the bed, where there was a dresser. He stretched and turned around and got the fright of his life.

"The shock was electric," Chris recalled. "Because I was still lying in the bed sleeping, and it was very clearly me lying there sleeping, my first thought was that I had died. I'm dead and this is the first step. I was just gasping. My head was spinning, trying to get a grip on things."

And then the phone rang.

"I don't know why, but I picked up the phone and said, 'Hello.' It was David. I immediately recognized his voice. I was overwhelmed, but at the same time I had this incredible sensation of joy." But David didn't stay on the line for long. "He told me that he didn't have much time and he just wanted me to know that he was all right, and to tell the rest of the family, then he hung up," Chris said.

"And then there was this enormous sucking sensation," said Chris, making a long, drawn-out slurping sound. "I felt like I was dragged, almost thrown, back into the bed, smack into myself." He woke up screaming. His wife, Sonia, who was asleep next to him, woke up to find a hysterical Chris.

"I was totally freaked out, I was shaking all over, I was sweating, my heart was beating like a racehorse's," said Chris.

Chris grew up in a scientific household. His father is a renowned nuclear physicist. Chris's upbringing was at odds with this experience. "My heart tells me that David was letting me know that he was OK. I really believed at the time that he was somehow communicating with me from beyond death," Chris said. "But my intellectual side says that's just silly. But it's so hard to rationalize; the experience was so real."

● ● ○

What Chris experienced was a particularly intense doppelgänger effect, also known in neuroscientific jargon as heautoscopy. It is different from an out-of-body experience in many ways.

In an OBE, the self, or center of awareness, gets dissociated from the physical body. The self identifies with a different location in space and has an altered perspective. The physical body itself is usually perceived as lifeless.

In heautoscopy, you perceive an illusory body, and your center of awareness can shift from within the physical body to the illusory body and back—there's self-location and self-identification with a volume in space, whether that volume is centered on the physical body or the illusory body. The perspective also shifts accordingly. In Chris's case, he was situated in his illusory body and then got sucked back into the physical body. But in other cases, such as for Brugger's young patient, one might experience this shift many times before the hallucination ends.

The other key components of heautoscopy are the presence of intense emotions and the involvement of the sensory-motor system. "Usually, the double is moving and there is interaction, there is sharing of emotions, of thoughts, and that's what's giving the impression of a doppelgänger," said neurologist Lukas Heydrich, who was at the Swiss Federal Institute of Technology in Lausanne when I met him.

To understand the differences in the neural activity associated with (or the neural correlates of) just seeing a visual double while remaining anchored to the physical body versus actually interacting and switching perspectives with a double, Heydrich and Blanke decided to study patients with brain damage who also experienced these autoscopic phenomena. In 2013, they published the results of the largest such sample to date. The data tell us a lot about the neural correlates of such experiences.

Patients who reported autoscopic hallucinations had lesions in the occipital cortex. Heydrich and Blanke hypothesize that simply seeing a double is not a disturbance of the bodily self, since self-identification, self-location, and first-person perspective remain intact. Rather, the hallucination is the result of the loss of integration between visual and somatosensory signals.

Patients who reported heautoscopic hallucinations, on the other hand, showed damage to the left posterior insula and adjacent cortical areas. Given that heautoscopic hallucinations involve emotions, it's revealing that the insular cortex is implicated. We saw how in depersonalization, lowered activity in the insula was correlated with the symptoms of emotional numbness (recall the way tattooed Nicholas back in Nova Scotia felt a lack of emotional vividness). The insula is the hub that integrates visual, auditory, sensory, motor, proprioceptive, and vestibular signals with signals from the viscera. It's the brain region where the body's states seem to be represented and the representations are eventually manifested as subjective feelings.

Heydrich and Blanke hypothesize that disturbances in the integration of signals in the insula are leading to the doppelgänger effect. If everything is working as it should, the insular cortex, particularly the anterior part of the insula, is thought to create a subjective feeling of one's body—a perception that includes emotions and actions. When abnormalities arise in the integration, it's as if there are now two representations of the body instead of one, and somehow the brain has to choose the representation in which to anchor the self, or rather choose which representation to imbue with self-location, self-identification, and first-person perspective. The hallucination happens when that trinity of parameters defining the basic bodily self switch between different body representations, one of which is not centered in the physical body in terms of geometric coordinates.

Metzinger and Blanke believe that these disturbances of the bodily self are helping them identify the basic attributes you need in order to feel like an embodied self—what they call the *minimal phenomenal self*. To start with, they argue that the sense of agency is not key to a minimal phenomenal self, since you can create a sense of being a body in some other location by merely passively stroking someone's back and messing with their visual input. This requires no agency on the subject's part. "From a philosopher's point of view, it is important to find out what is necessary and what is sufficient for self-consciousness," Metzinger told me. "We have shown that something that most people think is necessary is not necessary, namely agency."

Rather, the minimal phenomenal self is a more primitive embodied self. Metzinger argues that this feeling of being embodied is a prereflective, prelinguistic form of selfhood—something that comes long before we have the capacity to use the personal pronoun in phrases like "I think." There's no narrative here, just the organism having the sense of being a body. The next step in the process is when this primitive selfhood, which is merely an embodiment, turns into selfhood as subjectivity. "If you not only feel that you are in that body, but if you can control your attention, and attend to the body, that's a stronger form of selfhood," said Metzinger. "Then you are something that has a perspective, something that is directed at the world, and something that can be directed at itself. That is more than mere embodiment."

We are now getting close to the heart of the debate over the self. The issue that concerns philosophers and neuroscientists is the subjectivity of the self. Where does that come from? As you can imagine, opinions differ. Blanke, for instance, disagrees with Metzinger's idea that attention is needed for a strong, subjective selfhood. Blanke thinks that selfhood that arises out of a combination of a sense of body ownership, self-location, and first-person perspective should be independent of attention. We don't have the empirical data to sort out these nuances. Still, despite these disagreements, there's excitement that studying autoscopic phenomena will get us closer to understanding the "I," the self-as-subject, than almost anything else.

Why did this minimal phenomenal self evolve in the first place? Most likely as an adaptation that let the organism orient itself and function better in its environment. If the brain evolved to help the body avoid surprises and remain in homeostatic equilibrium and to effectively move around in its environment, then representing the body in the brain was a necessary step to fine-tune these abilities. Eventually, this representation became conscious, further enabling the organism to be aware of the body's strengths and weaknesses, which must have given it a survival advantage. But in this case, rather than physical attributes, it was the self that was being honed in evolutionary time.

● ● ○

Saying that the brain models the body doesn't quite get at the heart of the sense of ownership of the body, or the sense of *mineness.* The brain models things in the environment too, but they don't have the same feelings attached to them. Take the rubber hand. Once the illusion sets in, you feel as if the rubber hand is yours, but before the illusion, the rubber hand does not have that feel of *mineness* to it. We saw in the chapter on BIID that Metzinger's phenomenal self model (PSM) offers one kind of "representationalist" explanation. If the rubber hand is in the world-model constructed by the brain, it does not have a feeling of *mineness*, but if it's incorporated into the PSM, it becomes mine.

There are mechanistic explanations for the feeling of *mineness.* We saw hints of this in exploring schizophrenia. The feeling of agency—that *I* am the initiator of my actions, or a feeling of *mineness* to one's actions—may be the outcome of the brain being able to predict the consequences of one's motor actions correctly. If something goes wrong either in the prediction phase, when the prediction is being compared to the actual outcome of the action, or for that matter anywhere in that pathway, then an action may not have the feeling of being self-initiated. And so it's implicitly attributed to an external agent—to non-self.

Could the feelings of body ownership arise due to similar mechanisms? Philosopher Jakob Hohwy has argued that the phenomenon of *mineness* in general—whether for actions or perceptions—can be the outcome of a predictive brain. So, in this way of thinking, the brain is using its internal models to predict the causes for various sensory signals, and the brain's job is to minimize prediction errors. So, just like a sense of agency results from successful predictions, a sense of body ownership would also result from minimal prediction errors for the body as a whole.

● ● ○

Given all the talk of the minimal self and the extended narrative self, it's easy to get misled into imagining the self as an onion that can be peeled layer by layer, or as an orange that can be segmented. Yes, it's true that our narrative self has, in an evolutionary-biology sense, evolved after the bodily or minimal self, but in the complex selves that we are today, modern neuroscience is clearly telling us that the bodily self informs the narrative and your narrative can change how your body feels, and both the bodily self and one's narrative are influenced by one's cultural context. In this emerging understanding, brain, body, mind, self, and society are inseparable, insofar as a functioning human being is concerned.

Are there ways to test some of these linkages? Does an out-of-body experience influence perception and the construction of the narrative self?

Ehrsson's team had people experience a full-body illusion in which they felt ownership of virtual bodies that were as small as Barbie dolls (about a foot high) or as big as a thirteen-foot-high giant. Then they were asked about the objects they were perceiving (cubes of different sizes, placed at a constant distance from the camera). The subjects were more likely to perceive objects as being larger and farther away when they identified with Barbie dolls, and smaller and nearer when they felt they were giants. "One's own body size serves as an approximate reference for the entire external world in view," the team concluded. This is good evidence for the primacy of embodiment for our sense of self.

Ehrsson's lab also tested the effects of an out-of-body experience on episodic memory with an elaborate setup. They induced full-body illusions in subjects using the usual complement of head-mounted displays and synchronous stroking. During the illusion, subjects felt like they were watching a scene in a room from a different location than the location of their physical body. In the scene, an actor played the part of a professor and interacted with the subjects (all of whom were university students). The actor used a script that had been adapted from a Harold Pinter play called *One for the Road* (the adapted script was "not so dark and heavy as the original," said Ehrsson), and the interaction involved an oral examination, in which the student responded to questions. What Ehrsson's team wanted to answer was this: did people remember the episodes any less when they were under the influence of an illusion of being outside their physical body? In other words, does the brain's ability to encode episodic memory (which is essential to our narrative self, as we saw with Clare's father and Allan in chapter 2) depend on our being embodied in the physical body?

The short answer is yes. Those subjects who were out-of-body during the encounter with the professor were less able to recall the episodes, compared to those who were in-body. "The out-of-body-created memories were significantly less structured in terms of temporal and spatial order of events, and less vivid," Ehrsson told me in an email.

If this is so, how, then, does one make sense of the vivid recollections of people who have had out-of-body and heautoscopic experiences? "Those memories are probably less vivid and temporally structured (more fragmented and less coherent) than they would have been if the same event would have been experienced in-body," said Ehrsson. At least initially. Then, by repeated retelling of their experience, people consolidate their fragmented memories and are eventually able to recall and narrate the experience with considerable vividness. It's also possible that the dramatic and emotional nature of such experiences counters some of the out-of-body-induced memory impairment. Regardless, the basic embodied self seems fundamental to the more evolved, cognitive, narrative self in more ways than one.

However, in none of these conditions we have explored so far—whether in labs or in the subjective experiences of people—does the narrative self ever shut down fully. It does happen, sadly, in Alzheimer's disease, but other cognitive abilities deteriorate too, debilitating the person in the process. But what if there were a way to be just the bodily self—just the organism living in the present moment, sensing, feeling, without the chatter of the narrative self? It almost sounds mystical, even New-Agey. But that's where we are headed.

8

BEING NO ONE, HERE AND NOW

ECSTATIC EPILEPSY AND THE UNBOUNDED SELF

> If the doors of perception were cleansed every thing would appear to man as it is, infinite.
>
> **—William Blake**

> I feel a happiness unthinkable in the normal state and unimaginable for anyone who hasn't experienced it . . . I am then in perfect harmony with myself and the entire universe.
>
> **—Fyodor Dostoevsky**

Zachary Ernst was eighteen, in his second semester of college at Western Michigan University in Kalamazoo, when he had his first epileptic seizure. It was winter, a time when Kalamazoo is usually cold, dark, and cloudy. Zach and his girlfriend were sitting in his dorm room when he suddenly felt panicked. His mood darkened, suicidally so. He began hearing music that clearly wasn't playing anywhere, except in his head. A terrified Zach made his girlfriend take him to her parents' place nearby, which she did reluctantly. The whole episode left him drained of energy. Convinced it was just a panic attack, Zach ignored it, hoping it wouldn't happen again. It did, again and again, almost every day.

The seizures were so exhausting that Zach could not even summon the strength to see a doctor. Eventually, during a lull in the spate of seizures, he felt well enough to see a physician, who sent him to a psychiatrist, who immediately suggested he see a neurologist. When an EEG and an MRI revealed nothing, the neurologist started him on Tegretol, an anticonvulsant drug. But the seizures kept coming, sometimes two or three a day. The neurologist kept increasing the dosage, until Zach was taking 1,000 milligrams of Tegretol a day. "Years later I got a second opinion and they were horrified at what I was taking," Zach told me. "And they actually hospitalized me to take me off the medication . . . slowly."

But in the intervening years between the first diagnosis and the second opinion, Zach's condition worsened. His short-term memory failed him badly (a side effect of the Tegretol, it would turn out). He had entered college to study math, but he could barely remember when or where his classes were. He had to carry a schedule with him at all times. Until the seizures began, he had been tackling upper-level calculus, non-Euclidean geometry, and group theory with relative ease. But he began to find the math tests increasingly difficult. Oddly, he was able to handle philosophy better—the grades were based not on classroom tests but papers, which he could write in his room while referring to his notes. He didn't have to rely on his memory. "I was failing all my math classes and getting A's in all my philosophy classes," he said.

The seizures continued. When they came, they made him lethargic; he found it difficult to speak or walk. He learned to recognize an imminent seizure. He'd walk over to the original part of campus, one of the few places in Kalamazoo with older buildings, to wait out the episode. "This very crushing feeling of sadness would wash over me, to the point where I'd have certainly tried to kill myself if I had the energy to do it," he said. "It was very, very severe and very sudden. It would go away as quickly as it came."

Given the overwhelmingly negative emotional tone of these seizures, it's no wonder that Zach at first didn't realize that he was having another type of seizure also. They were less frequent, but they too came unbidden. And they were pleasant. Very pleasant. He might have had them as a kid, but he can best remember the ones that happened during his college years. The world around him would turn sharp and vivid, as if until then he had been seeing everything on a flat screen and suddenly someone had taken the screen away to expose a 3-D world. He noticed details in ways that he wouldn't otherwise. "If I saw a tree, it would be like seeing a real tree after you had only seen pictures of a tree," he told me. "You could take in all of the details of the whole tree all at once and see the textures of everything. It was very, very beautiful."

Time seemed to slow down. He would be walking down a city block at his normal pace and what normally would take a few minutes would feel like it had lasted an hour. "It felt like time was being stretched out," he said. "Like you were experiencing more every second than you normally do." Put another way, Zach was living in the moment. "There was certainly no place else to be at that time," he said. "I was very focused on where I was [at] that exact time. That's very pleasant. You don't worry about things that are going to happen in an hour, a year."

Was it an unusual state for him, living in the moment? I asked.

"Oh, it's unusual for me," he said, laughing. "I'm normally drifting around everywhere."

Even more than the vividness and the slowing of time, the feeling that left the most indelible mark on Zach was a pervasive sense of certitude. "The world sort of looked like a very well composed photograph or a very well composed painting, where the objects are placed in just the right way that it brings out an aesthetic," he recalled. "And Kalamazoo is not a pretty town. It's very gray; it's very depressing, in my opinion. That was certainly not how I normally thought about it."

There was also the feeling of clarity, of knowing. "I felt like I knew everything about my environment so directly that there was no inference going on. It's some strange feeling of certainty that the way the world is, is exactly how it should be, and how it was arranged to be," he recalled. "It just cried out for an explanation. Very ordinary objects—tables, and chairs, and trees, everything—felt so powerfully like they are laid out with such precision and intentionality. I got an overwhelming feeling that there was an agency behind it."

Listening to Zach, it was hard not to think of mystics and descriptions of their otherworldly experiences. I said so. He agreed. Having grown up an atheist, Zach didn't equate his experiences with the existence of anything supernatural. "But it seems really clear to me that what mystics are describing is this," he said.

As for himself, he remains an atheist. He became an associate professor of philosophy at the University of Missouri–Columbia—a profession that reinforced his skepticism. But he was at pains to point out that his post-seizure views cannot negate the "truth" of the seizures when he was in their throes. "During the seizures, it was impossible to doubt that there was a kind of agency behind the world. That was not open to debate, so to speak," he said. "It was such an immediate belief; there was no way to prevent myself from having it."

● ● ○

Fyodor Dostoevsky would have agreed. The Russian novelist is one of the best-known literary figures who suffered from epilepsy. While his seizures often left him feeling a dark dread ("as if I had lost the most precious being in the world, as if I had buried someone," Dostoevsky told his wife, Anna), historians have identified instances when Dostoevsky talked of soaring high, just moments before being knocked unconscious by his seizure. "A happiness unthinkable in the normal state and unimaginable for anyone who hasn't experienced it . . . I am then in perfect harmony with myself and the entire universe," he told his biographer Nikolay Strakhov of those moments. "The sensation is so strong and so pleasant that one would give ten years of life, perhaps even one's whole life in exchange for a few seconds of such felicity."

Many of Dostoevsky's fictional characters suffered from epilepsy. Prince Myshkin, the protagonist in *The Idiot*, has ecstatic auras at the start of his fits: "a moment or two when his whole heart, and mind, and body seemed to wake up to vigour and light; when he became filled with joy and hope, and all his anxieties seemed to be swept away for ever." Myshkin even tells the novel's villainous character Rogozhin, "I feel then as if I understood those amazing words—'There shall be no more time.'" But Myshkin is no fool. He realizes upon reflection that these unusual states are only due to his disease, a lower, not higher, form of existence. Still, he cannot shake off the truth of those moments. Why should it matter, Myshkin ponders, "if when I recall and analyze the moment, it seems to have been one of harmony and beauty in the highest degree—an instant of deepest sensation, overflowing with unbounded joy and rapture, ecstatic devotion, and completest life?"

Did Dostoevsky fabricate the accounts of his ecstatic "auras," genius novelist that he was? That's exactly what French neurologist Henri Gastaut eloquently argued in 1977, after systematically analyzing the available evidence: "I believe that the grand mal seizures of his epilepsy lacked any such ecstatic aura, but were preceded on rare occasions by mild alterations of consciousness which the author's originality of thought and literary genius led him to describe as a feeling of bliss."

It didn't take long, however, for this view to be overturned. In 1980, Italian neurologists reported the story of a thirty-year-old man who had been having ecstatic seizures since he was thirteen. He didn't see the need to go to a physician, but then he had a tonic-clonic, or grand mal, seizure; the physician referred him to the neurologists. Their account of his ecstatic seizures is edifying: "He says that the pleasure he feels is so intense that he cannot find its match in reality. . . . All disagreeable feelings, emotions, and thoughts are absent during the attacks. His mind, his whole being is pervaded by a sense of total bliss. . . . He insists that the only comparable pleasure is that conveyed by music. Sexual pleasure is completely different: once he happened to have an attack during sexual intercourse, which he carried on mechanically, being totally absorbed in his utterly mental enjoyment. The neurological examination was negative." The neurologists even managed to get an EEG recording while the young man was having a seizure, which he claimed created a feeling of ecstasy. Based on those recordings, the neurologists concluded that seizures in the temporal lobe can make people feel ecstatic.

That's where things stood until recently. Fabienne Picard, a neurologist at the University Hospital Geneva in Switzerland, stumbled upon Dostoevsky's writings on epilepsy and his epileptic characters while conceiving and producing her documentary *Art & Epilepsy.* Until then, her work had focused mainly on nocturnal frontal lobe epilepsy, which as the name suggests causes seizures mostly when the person is asleep. But after getting acquainted with Dostoevsky's ecstatic auras, she began paying greater attention to some of her other patients. "When they really explained their feelings, it was incredible," Picard told me. "It was very close to Dostoevsky's descriptions."

Epileptic seizures are broadly divided into two groups: generalized and focal. In generalized seizures, electrical discharges overwhelm the entire cortex, and can often lead to loss of consciousness. Ecstatic seizures are of the second kind: focal or partial, in which the electrical storm is confined to a small region of the brain, and the patient often remains conscious.

Detailed accounts of ecstatic seizures are scant in the medical literature. "These seizures are not so frequent, but I think that they are also probably underestimated, because people are sometimes reluctant to explain them," said Picard. "Because the emotions are so strong and strange, maybe they feel embarrassed to speak about them; maybe they think the doctor will find them mad." Also, given the blissful and pleasurable nature of such seizures, some patients may not even go to see a neurologist unless the seizures spread to other areas of the brain, causing loss of function or even loss of consciousness.

As Picard cajoled her patients to speak up about their ecstatic seizures, she found their sensations could be characterized using three broad categories of feelings. The first was *heightened self-awareness.* For example, a fifty-three-year-old female teacher told

Picard, "During the seizure it is as if I were very, very conscious, more aware, and the sensations, everything, seems bigger, overwhelming me." The second was a sense of *physical well-being.* A thirty-seven-year-old man described it as "a sensation of velvet, as if I were sheltered from anything negative." The third was *intense positive emotions,* best articulated by a sixty-four-year-old woman: "The immense joy that fills me is above physical sensations. It is a feeling of total presence, an absolute integration of myself, a feeling of unbelievable harmony of my whole body and myself with life, with the world, with the 'All,'" she said.

As far as Picard was concerned, these descriptions were pointing toward one brain region—the insular cortex—and to the work of Bud Craig, a neuroanatomist at the Barrow Neurological Institute in Phoenix, Arizona. In 2002, Craig published a remarkable paper in *Nature Reviews Neuroscience,* titled "How Do You Feel?" and followed it up with a paper in 2009 in the same journal, titled "How Do You Feel—Now?: The Anterior Insula and Human Awareness." In these papers, Craig brought a body of experimental work—his own and that of others—to bear on his hypothesis that the anterior insula is key to human awareness, maybe even the seat of the "sentient self."

We have seen the insula implicated in Cotard's syndrome, depersonalization disorder, and the doppelgänger effect—all of which involve distortions in one's perception of body states and emotions. The insula is deep in the brain, buried inside the lateral sulcus, the fissure that divides the frontal and parietal lobes from the temporal lobe. Its main function seems to be to integrate information about the internal state of the body with external sensations. There's also evidence that the processing of these signals gets progressively more sophisticated as one moves from the posterior to the anterior part of the insula.

While the posterior insula represents objective properties, such as body temperature, the anterior insula produces subjective feelings of body states and emotions, both good and bad: the anterior insula could be responsible for creating the feeling of "being."

Picard was intrigued by Craig's hypothesis. Her patients' explanations of how they felt during their ecstatic seizures suggested that the symptoms may have to do with a dysfunctional insula, specifically the anterior insula. And one of her patients provided preliminary evidence that this could be the case. The patient, the thirty-seven-year-old man who described his ecstatic aura as a "sensation of velvet," had been operated on in 1996 for a tumor in the right temporal region. He was free of seizures until 2002, when they began again, but less frequently. While closely monitoring him during one of his seizures, neurologists managed to get a SPECT image of his brain. This involves injecting a nuclear tracer as the patient is having a seizure (something the patient agrees to in advance). In about thirty seconds, this tracer is absorbed by brain regions with greater blood flow, a proxy for greater brain activity. Thirty minutes later, the patient is wheeled into a scanner and brain scans can reveal the region that was most active during the seizure. In this man's case, it was the right anterior insula.

Two other patients enriched Picard's understanding. One was Zachary Ernst, who provided Picard with insightful descriptions of his ecstatic seizures. The other was a seventeen-year-old farmer from Romont, Switzerland, who came to Geneva to be closely monitored during his seizures. Picard asked him for a detailed explanation of his experiences, and he wrote a report for her, in which he mentioned what he called his "absences." There were big absences, with loss of consciousness. And there were small ones, when he remained aware, and time seemed to slow down: those around him at the time would tell him that his seizure had lasted one or two seconds, but he felt immersed in them for longer, much longer. It was hard for him to tell how long. And these seizures would be triggered by something pleasant in his environment, "a nice car going by, pictures, a color, a flower, a landscape, animals grazing, a bird singing, branches moving in the wind, a person smiling, a beautiful woman, a kiss, a caress, a thought about someone, a hope . . ."

Picard's neurological examination of the teenage farmer would further implicate the anterior insula as the source of such magical feelings.

● ● ○

The train from Geneva to Romont sped along the northern shore of Lake Geneva (or Lac Léman in French), until it reached Lausanne, and soon after, it veered away from the lake into mountainous terrain. When I got off at Romont, I was met by Catherine. It was her son Albéric, the young farmer, who had written a report for Picard. Catherine drove me to a farm to meet Albéric. During that short drive, she told me that Albéric was the third of six children and that though he had been the heaviest at birth, his delivery had been the most harmonious. It had been a water birth, as with all her children. And she had named him Albéric (Celtic for "king of bears") because he had been "very strong in [her] body . . . very virile."

Albéric had been an uncomplicated, happy kid. He loved nature (which, I'd discover, was in staggering abundance around their three-hundred-year-old farmhouse near Romont), walked around barefoot, loved the company of their cows, and by age three was clambering onto his godfather's tractor. He also began speaking at age three, much later than his sisters.

We met Albéric at the farm where he was working. The images I had built up of a cherubic kid vanished; here was a strapping nineteen-year-old, slightly disheveled from working on the farm, with a gentle, easy smile. He had cut himself while working; his finger was bleeding. But he nonchalantly got into the car and we drove back to Romont to talk about his epilepsy; he spoke in French and his mother translated.

The first seizure came when he was fifteen. Just one week before it happened, Albéric had an experience that subtly echoed elements of my first meeting with him. He and his godfather were at a mountain farm, some five thousand feet in altitude, when his godfather badly cut three of his fingers while using a woodworking machine. He asked Albéric to stay at the farm and care for the cows, which were being specially tended to before they had their first calves, and he drove off to the hospital. Albéric called his mother, and she said, "You shouldn't have let him go." When she arrived at the farm, Catherine found Albéric crying and in shock. He was traumatized, by the accident and because he had let his godfather drive away alone. "One week later it began, the seizures," Catherine told me.

It happened at the same mountain chalet. This time Albéric was with his father. They had finished cleaning the stalls for the cows and were sitting by the fireplace. Suddenly he felt a strange, unfamiliar taste in his mouth. Then he lost consciousness and soon thereafter the convulsions began, which he could not recollect. It was the first of his big "absences." His parents, unaware of what exactly had happened, brought him back down to their farmhouse. At home, Catherine realized that Albéric was still confused. She was bathing him with running hot water when he remarked at how strange it was that the chalet had hot water (up in the mountains they had to boil water). "He didn't realize he was home," recalled Catherine. That night his father slept near him. "We thought, He'll be never the same. . . . Something is broken in his person," said Catherine.

Albéric's big seizures continued, mainly at night, signs of which would be evident in the morning, in the form of extreme tiredness or even a bitten tongue. But he began having smaller seizures too, during which he remained aware. These he found to be an entirely different experience. The triggers were usually something pleasant, such as—unsurprisingly for a farmer—the sight of a tractor during harvest season. Nature had a great effect too. At times, he would feel as if the cows were speaking to him just before the seizure. The seizures themselves were pleasant as well. "He told me once it is just like a drug, these seizures," said Catherine. "When he doesn't have any, he waits, thinking it'll come. He thinks perhaps he'll never [be able to] live without them."

Unfortunately, Albéric continued to suffer from the more dangerous absences. One came when he was working as an apprentice at a farm. He woke up at four a.m. and walked over to his boss's house—wearing only his underwear. The boss asked, "What are you doing here?" and Albéric left. Soon, his boss noticed a light in the barn, so he walked over and found Albéric clambering barefoot onto a combine harvester. The key was in the machine. Albéric does not remember the incident.

Around the time he turned seventeen, he became Picard's patient. An MRI scan showed a benign tumor in the right temporal pole. When I went to see Picard in Geneva in March 2013, she showed me the room where Albéric had been monitored during one of his seizures. A technician sat in a room with monitors, which showed the EEG signals from patients in each of four rooms. Each screen was filled with rows and rows of squiggles—each squiggle a signal from one EEG electrode. They looked like the tracings of a seismograph. Only a trained neurologist or technician could spot the signature of a seizure. For instance, a focal seizure would generate spikes in one electrode or in a cluster of nearby electrodes. Inset among those electrical waveforms on each screen was the feed from a camera that was trained on the patient, lying in bed with an EEG skullcap covering his head. Albéric had been in one of those rooms when he had an eighty-second seizure. The EEG showed the seizure beginning in the right anterior temporal region. Upon onset of the seizure, Albéric was injected with a nuclear tracer. And subsequent SPECT imaging revealed that the seizure had involved increased blood flow (or activity) in the right insula, and also close to the tumor. Neurosurgeons operated on Albéric and removed his tumor.

Before his surgery, Albéric had to see a psychiatrist, because there was the danger of depression after brain surgery. Unflustered by the thought of depression, Albéric had told the doctors, "No problem; when it's not good, *cartouche*!" In other words, there was always the choice of ending one's life with a gun if things took a turn for the worse. He was joking, of course. "They were very shocked," recalled Catherine. But she knew what her son meant and had clarified that he was joking. "I was laughing. It's the way of thinking of the farmer; if a cow is not good, they shoot it. In the farm, it's totally normal to speak so. When we live in nature, we see this in life."

After his surgery, Albéric's condition improved for a while, but the more serious seizures returned, especially at night. The seizures are potentially dangerous when he's working on the farm, especially if one comes when he's on a tractor or a combine harvester. The family's response to his situation is typically earthy. They are researching the use of a dog that could alert Albéric of an imminent seizure. And he might undergo another surgery. Whatever the eventuality, "he'll have to live with it," says Catherine. "We'll be always his parents. We will always be here for him. That's no discussion."

Sadly, the one part of his epilepsy that he cherished—the ecstatic seizures—ceased after his surgery. Even so, in the brief period when Albéric allowed himself to be studied, Picard's intuition about the role of the insula in ecstatic seizures received strong support. Albéric's descriptions corroborated those of Zachary Ernst. There is something about an ecstatic seizure that changes one's self-awareness and relationship to the world. Picard wrote this in her case report about Albéric: "He felt a deepening awareness of the situation or conversation going on around him, a sudden clarity. It was as if he clearly understood everything, especially if he happened to be in the midst of a discussion with several people. He grasped it all simultaneously. Things suddenly seemed self-evident, almost predictable (yet without the feeling of knowing the future)."

Again, the strains of a mystical experience: the slowing of time; hyperawareness of one's environment; and a sense of certainty that all that's being perceived is as it should be. Another of Picard's patients that I met in Geneva, a forty-one-year-old Spanish architect, had this to say about her episodes of ecstatic seizures: "You are just feeling energy and all your senses. You take in everything that is around; you get a fusion. You forget yourself."

Picard acknowledged the seeming paradox here: the seizures made the person both extremely self-aware while simultaneously blurring the boundaries between one's self and the world, the result of which, as Dostoevsky wrote, is to be "in perfect harmony with myself and the entire universe," a sense of oneness with everything.

During our conversation in March 2013, Picard said she couldn't shake off the intuition that the insula is involved in these extraordinary experiences. "I am more and more convinced that there is something affecting the insula," she said. "But I don't have proof for any patient." The only bits of evidence she had were the nuclear-imaging studies, but such studies are not precise enough to pinpoint the brain regions involved in the seizures. A seizure is a dynamic, rapidly evolving neural process, and the tracer can take up to thirty seconds to become "fixed" in the brain—enough of a lag to cause a blurry image; it's as if you are trying to photograph a fast-moving car, but your camera's shutter speed is too slow. Picard wanted something unambiguous.

She got news of one such study the very next day. I was in her office when she received an email from Fabrice Bartolomei, a neurologist at the Hôpital de la Timone in Marseille, France. Bartolomei's surgical team had implanted electrodes deep inside the brain of a young woman suffering from ecstatic seizures. His email to Picard read, "We have explored the patient . . . The stimulations in the anterior insula trigger a pleasant sensation of floating and chills."

Picard shot off a reply: "I'm so happy!"

The connection between ecstatic epilepsy and Bud Craig's hypothesis of the insula as the seat of the sentient self had just become stronger.

● ● ○

"I don't believe the brain is a mystical place," said Craig in a talk he gave in Sweden in October 2009. "When René Descartes came to Sweden three hundred years ago and taught that humans know that they exist because they think . . . he left our brains somewhere off in this metaphysical space, but they really belong in our bodies, because that's what we are. We are biological organisms and our bodies are what our brains are designed to take care of."

As we have already seen, the brain takes care of the body by maintaining homeostasis, which involves keeping the body's physiology at an optimal state despite wide variations in the external environment. It was a detailed examination of one of the neural pathways involved in homeostasis—having to do with thermal regulation—that led Craig to the anterior insula.

In his talk in Sweden, Craig mentioned a paradox that had bothered him when he was a graduate student in the 1970s. The neuroscience textbooks he was reading talked about how pain and temperature are represented in the somatosensory cortex, the part of the brain responsible for the sense of touch. As we saw in chapter 3, the somatosensory cortex was mapped out in the mid-twentieth century by Wilder Penfield, and showed the relationship between cortical areas and the tactile sensations in various parts of the body: stimulating a specific part of the somatosensory cortex would cause in the subject a feeling of being touched in a specific part of the body. But not so for pain or temperature. "Stimulation of the somatosensory cortex almost never causes pain or a feeling of temperature, and a lesion of the somatosensory cortex almost never affects pain or temperature," Craig said in his talk. "I didn't understand why textbooks would contain a contradiction like this. Of course, I passed all my tests by saying this is the way it is." The question remained: where in the brain was the region dealing with pain and temperature?

As a neuroanatomist, Craig grappled with this problem. There were clues. To start with, there's a curious illusion, a staple of science museums worldwide, called the thermal grill (discovered in 1896 by a Swedish physician). The grill, as the name suggests, is a set of metal bars, alternately warm or cool. Neither the warmth nor the coolness is above the threshold of what's considered hot or cold enough to cause pain. However, place your hand on this grill and you'll likely experience a burning pain. "The thermal grill reveals a fundamental feature of the organization of the nervous system—in this case, a fundamental interaction between the feelings of pain and temperature," writes Craig.

In the mid-1990s, Craig and his colleagues used PET scans to study the brains of people while they were experiencing the burning pain induced by the thermal-grill illusion. The subjects were also studied when they touched the warm and cold parts of the grill one at a time, which resulted in no pain. Their findings were revelatory: the experience of pain was correlated with activity in the anterior cingulate cortex (ACC), whereas the mid-to-anterior insula was activated at all times (whether the thermal stimuli were painful or not).

In his subsequent studies also using PET scans, Craig showed that the posterior insula is responsible for representing temperature objectively, but the activity in the anterior insula was correlated not with the objective temperature but with the subjective perception of it. This is an interesting and crucial difference. Say you drink a glass of cold water. Seen from the perspective of Craig's results, the posterior insula is representing the actual temperature of the water, but depending on whether you drink the water on a hot day or an icy-cold day, your subjective feelings about the glass of water will differ—possibly from an extremely pleasant sensation to something that is undesirable. This subjective feeling is what's being represented in the anterior insula. And his work on the thermal-grill illusion suggested that when the sensation goes from being merely pleasant or unpleasant to thermal distress (something that the body has to act upon), then both the anterior insula and the anterior cingulate cortex are activated.

This has led Craig to argue that a feeling is more than just a perception of a body state; it also includes the motivation to do something about it. As Craig puts it, "Activation of the ACC is associated with motivation, and activation of the insula is associated with feeling, which together form an emotion." And the emotion drives homeostasis—if being out in cold weather becomes painful, the pain drives the organism toward seeking warmth.

And so it was that the study of pain and temperature led Craig to the insula, and his idea that this deep-brain region is crucial for self-awareness. A body of work is now showing that the anterior insula and anterior cingulate cortex are activated by a whole host of feelings, from anger to lust, from hunger to thirst. Craig rolled his work and that of others into a compelling hypothesis. He argued that the anterior insula is the brain region responsible for our feelings—the neural substrate for the subjective awareness of our body's physiological state. It involves the integration of external sensations, internal sensations, and states representing the body's motivation for action. "It seems to provide the anatomical basis for emotional awareness."

Craig argues that the anterior insula provides the grounding for the "material me" or the self-as-object, creating the moment-to-moment mental image of "the material self as a feeling (sentient) entity." And since much of the material self is based on an unchanging body (at least on short timescales), it could be the "source of the sense of continuous being that anchors the mental self." As Craig told me during a phone interview, "The immediate self, present in this moment, is based in the anterior insula."

It was these studies that led Picard to her hypothesis that the anterior insula could be the focus of ecstatic seizures. Were such seizures intensifying the experience of being the material me, the self as experienced here and now? The best evidence for her idea came when Fabrice Bartolomei emailed her saying that they had induced feelings associated with ecstatic seizures by stimulating a patient's anterior insula directly.

● ● ○

Fabrice Bartolomei's patient was a twenty-three-year-old woman. She first came to see Bartolomei along with her boyfriend, who was curiously suspicious of Bartolomei. "There was tension during the consultation," Bartolomei told me during our phone conversation. Still, he examined the woman. She had started having seizures at fifteen, and stopped going to school as a result. She had a difficult personality, with aggressive, sociopathic tendencies. She was oppositional and moody during the consultations; her boyfriend came along to most of them at her insistence, and his negativity didn't help matters either. Despite all this, her symptoms had a silver lining. Her seizures always began with moments of ecstasy—much like Prince Myshkin's—before the seizures eventually knocked her unconscious.

"[Given] the mood of the patient was not very good, I was a bit surprised that the beginning of the seizures could [induce a] sensation of floating, with a strong shivering," said Bartolomei. The patient reported feeling happy during the ecstatic aura at the start of her seizures. "There was a sort of contrast between the sensations during the seizure onset and the general behavior of the patient."

The young woman had come to see Bartolomei because her epilepsy was resistant to drugs, and scalp EEG was not enough to locate the origin of her seizures. Bartolomei decided to insert depth electrodes into the patient's brain, to record brain activity during seizures and home in on the epileptogenic tissue, which could then be surgically excised. Bartolomei's measurements suggested that the seizure first began in the temporal lobe but quickly spread to the anterior insula in less than one second—supporting the idea that this region was triggering the blissful feelings at the beginning of the seizure.

When Bartolomei used the same electrodes to stimulate his patient's brain in specific places, one by one, the patient initially turned aggressive—she was reacting to the procedure, which Bartolomei accepts can be difficult for patients. This makes the following sequence of events all the more remarkable. Of the first eight electrodes, only the one stimulating the amygdala induced a response—in this case an unpleasant feeling (in addition to the patient's antagonism toward the procedure in general). But when the electrode in the anterior insula was activated, things changed. "The first thing I saw was a change in the facial expression. She seemed to be more happy, and less in tension," Bartolomei told me. The patient reported feelings that were akin to what she felt during her ecstatic auras. "I feel really well with a very pleasant funny sensation of floating and a sweet shiver in my arms," she told the doctors. And the greater the intensity of the stimulation, the greater was the "funny sensation." Bartolomei cautions that this is just one case, yet the evidence is strongly suggestive of the insula's involvement in ecstatic seizures. "The only site where we obtained this kind of pleasant sensation was the anterior insula," he told me. "We did not obtain this by the stimulation of the temporal pole, the amygdala, or the hippocampus."

After the exploratory stimulations, Bartolomei suggested surgery to his patient to remove the epileptogenic tissue, but thus far she has decided against it. Nonetheless, this patient's experiences have given Picard the much-needed "proof" for the role of the anterior insula in ecstatic seizures. Picard is more and more convinced that hyperactivity in the anterior insula is causing these feelings of bliss, well-being, and heightened self-awareness.

Neuroscientist Anil Seth, the researcher from the University of Sussex who has hypothesized that the brain's predictive mechanisms are involved not just in the perception of external stimuli but also of internal body states, is impressed by this work. "The fact that the direct electrical stimulation of the insula does elicit these kinds of feelings is pretty compelling," he said. The evidence also is consistent with findings that show that the insula is underactive in people with depersonalization disorder, in which they "describe the world as being drained of sensory and perceptual reality," said Seth. A hyperactive insula during ecstatic seizures produces the opposite effect.

● ● ○

"One bright May morning, I swallowed four-tenths of a gram of mescalin dissolved in half a glass of water and sat down to wait for the results." Thus began Aldous Huxley's extraordinary adventure in the spring of 1953, documented in his book *The Doors of Perception*. Huxley took the drug mescaline under the supervision of psychiatrist Humphry Osmond (who reportedly did not "relish the possibility, however remote, of finding a small but discreditable niche in literary history as the man who drove Aldous Huxley mad"). As it happened, Huxley did not go mad.

An arrangement of brightly hued flowers in a vase, which Huxley had found distasteful just hours before he popped the pill, had morphed in his perception. "At breakfast that morning I had been struck by the lively dissonance of its colours. But that was no longer the point. I was not looking now at an unusual flower arrangement. I was seeing what Adam had seen on the morning of his creation—the miracle, moment by moment, of naked existence." When asked if he found the bouquet agreeable or disagreeable, he replied it was neither. "It just is," he said.

He found his perception of space and time altered. "Space was still there; but it had lost its predominance. The mind was primarily concerned, not with measures and locations, but with being and meaning. And along with indifference to space there went an even more complete indifference to time. 'There seems to be plenty of it,' was all I would answer, when the investigator asked me to say what I felt about time. Plenty of it, but exactly how much was entirely irrelevant. . . . My actual experience had been, was still, of an indefinite duration or alternatively of a perpetual present."

Osmond would go on to coin the term "psychedelic," for the effect drugs such as mescaline, psilocybin, and LSD have on the mind ("To fathom Hell or soar angelic, just take a pinch of psychedelic," he would write to Huxley in response to a rhyme that Huxley had composed, as they attempted to describe the mind-bending nature of these drugs).

It's not surprising that accounts such as Huxley's and the descriptions of people having ecstatic seizures seem eerily similar. Neuroimaging studies of people taking psychedelic drugs such as psilocybin have also shown hyperactivity in the insular cortex and the anterior cingulate cortex. In one double-blind study of fifteen male subjects, researchers found that taking ayahuasca—a psychoactive tea used in shamanistic rituals in the Amazon—caused increased blood flow to the anterior insula, among other brain regions.

In both ecstatic seizures and the use of psychedelic drugs, one of the most intriguing effects is the change in the perception of time. Recall Prince Myshkin's words in *The Idiot:* "I feel then as if I understood those amazing words—'There shall be no more time' "; or Zachary Ernst's and Albéric's sense that time slowed down during their seizures. Bud Craig's model provides an explanation.

In Craig's model, the anterior insula integrates interoceptive, exteroceptive, and the body's state of action to create a "global emotional moment" once every 125 milliseconds. It's these global emotional moments strung together that give us a continuous sense of self, even though the moments themselves are discrete, he argues. It's like watching a movie—even though the cinema screen is displaying twenty-four discrete frames per second, we perceive a seamless continuum. A hyperactive anterior insula could potentially generate these global emotional moments faster and faster, leading to a subjective sense of time dilation. This is not unlike a high-speed camera shooting hundreds or thousands of frames per second—when played back at normal speed, we get to see everything in slow motion, as if time has slowed down. Craig also posits that the anterior insula might have a buffer that can hold a few such global emotional moments—a few that have just passed, the immediate present, and a few that are predicted to come. If you think of yourself as a series of such global emotional moments spanning decades, then the buffer is like a small window a few seconds wide. This is, of course, entirely unproven—but the idea gets to the heart of what philosophers are debating when it comes to those who say there is a self and those who say there isn't. For example, philosopher Dan Zahavi's notion of a minimal self necessitates the presence of a mental structure that can hold in place a few moments of subjective experience—past, present, and future—to construct the subject of experiences. Could the anterior insula be providing this mental structure? It's an intriguing speculation at this point.

If the anterior insula is predicting future states, then it helps explain yet another commonality between ecstatic auras and psychedelic experiences, which is the feeling of certainty, as if everything is the way it should be. This fits well with the predictive or Bayesian brain hypothesis, which we encountered in the context of autism and depersonalization disorder—the idea that our perceptions may be the brain's best guess as to the causes of sensations, something it needs to do in order to minimize surprise and maintain the body in homeostatic equilibrium.

Here, the hypothesis is that the insula is a key brain region involved in predicting the most likely cause of the various external and internal sensory signals it integrates. If the prediction error is small, we feel good; if it is large, we feel anxious. And anxiety is the brain's way of getting the body to respond—something is not quite right, and action is necessary. But like anything else, even this prediction error signal generator can go awry.

On one hand, it can lead to chronic anxiety or neuroticism. In 2006, Martin Paulus and Murray Stein argued that chronic anxiety is the result of a malfunctioning anterior insula, as it constantly generates higher than normal prediction errors. Picard posits that the opposite may be happening in ecstatic seizures. The electrical storm in the anterior insula may be disrupting the mechanism, resulting in few or no prediction errors. As a result, the person is left feeling as if nothing is wrong with the world, that everything makes sense, generating a feeling of absolute certainty.

Anil Seth agrees that this is a viable hypothesis. "In some ways the phenomenology of ecstatic seizures is the opposite of pathological anxiety," he says. "You have this sense of complete, peaceful certainty, [whereas] anxiety is a pathological visceral uncertainty about everything as reflected in bodily state."

It's uncanny how these feelings of serene certainty, heightened awareness, and a slowing of time also underpin accounts of mystical experiences. Picard's patients certainly attributed religious meaning to their seizures. "Some of my patients told me that though they are agnostic, they could understand that after such a seizure you can have faith, belief, because it has some spiritual component," she says. "Probably some people with mystical experiences actually had ecstatic seizures."

This brings us to the curious paradox about such experiences: the subject has a heightened self-awareness of oneself and one's environment while simultaneously feeling as if the boundary between oneself and the world has dissolved, leading to a feeling of oneness.

What's happening? In his book *Flow: The Psychology of Optimal Experience,* Mihaly Csikszentmihalyi provides some clues. Csikszentmihalyi defines flow as "joy, creativity, the process of total involvement with life." And flow encounters a similar paradox: the loss of self-consciousness. As Csikszentmihalyi puts it, "One item that disappears from awareness deserves special mention, because in normal life we spend so much time thinking about it: our own self. Here is a climber describing this aspect of the experience: 'It's a Zen feeling, like meditation or concentration. One thing you are after is the one-pointedness of mind. You can get your ego mixed up with climbing in all sorts of ways and it isn't necessarily enlightening. But when things become automatic, it's like an egoless thing, in a way.'"

Even though there is a loss of self-consciousness, Csikszentmihalyi adds, "the optimal experience involves a very active role for the self." That's the paradox. The climber, for example, is intensely aware of every aspect of his body and the mountain face, and yet is claiming the cessation of some aspect of his self. In Csikszentmihalyi's view the "loss of self-consciousness does not involve a loss of self, and certainly not a loss of consciousness, but rather, only a loss of consciousness *of* the self. What slips below the threshold of awareness is the *concept* of self, the information we use to represent to ourselves who we are."

So, it is the self that knows and obsesses about itself—the reflective, narrative, autobiographical self—that recedes, while the minimal, embodied self is fully present and engaged. "It is phenomenologically interesting that you can have these coexisting experiences of heightened self-awareness and heightened connection [with the world]," Seth told me. "To me that sort of implies that the experience of the division between the body and the world is perhaps more generally labile and flexible than we often assume."

A few millennia ago, a monk made the point that not only is the division between self and others flexible and labile but that there is no such thing as a self; if we were to go looking for a self that underpins our experience of "I," "me," and "mine," we'd find nothing, and it's our attachment to this false notion of an enduring self that's the cause of human suffering, This brings us to a place where I began my journeys for this book—Sarnath, India, where the Buddha, as the legend goes, gave his first sermons after he understood the nature of the self. We will end where it all begins, when we simply ask ourselves: "Who am I?"

EPILOGUE

The city of Varanasi gets its name from the rivers Varuna and Asi. Both rivers empty into the Ganga—India's longest and most sacred river. Varanasi's famous ghats—steps that descend to the riverbank—stretch along a crescent, from where the Varuna meets the Ganga on the northern end to where the Asi joins the river farther south. The city too meets the Ganga—the steps of its ghats leading pilgrims and ordinary folk alike down to the water's edge.

Near the confluence of the Varuna and the Ganga is a place called Rajghat. The Archeological Survey of India has, over decades, unearthed the remains of an ancient city in Rajghat, some of which date back to the sixth century BCE. Legend has it that around this time a monk, a former prince, crossed the Ganga, reached Rajghat, and then walked a distance of six miles or so to Sarnath, where he gave his first-ever sermon. The monk, who was in his mid-thirties, came to be called the Buddha.

The walk from Rajghat to Sarnath in the Buddha's time may have been idyllic. When I visited, it was monsoon season. The villagers advised me against walking the mud path. I took to the road in an auto-rickshaw. India spilled over onto the road from Rajghat to Sarnath—shops selling handwoven wicker baskets, terra-cotta urns and jars, stone tiles, and the obligatory government-licensed liquor. A young boy—about three or four years old—was trying to fly a kite, but the length of the string was hardly enough to get the kite off the ground. Somewhere midway through the journey, after we crossed the *purana* ("old" in Hindi) bridge, the road changed from asphalt to one paved with stones—making for a bone-jarring ride in the auto-rickshaw, its tiny wheels catching every gap between the not-so-closely-spaced stones. Fetid pools of rainwater lined the road; cars and buses squelched their way through, creating a muddy mess. I couldn't help feeling that the walk might have been better.

When the road reached Sarnath, everything quieted down. It was back to smoother asphalt. The trees lining the roads were old, and sat well amid the region's history. I felt an anticipation approaching a sacred place, but that was jarred by the sight of a garish temple festooned with bright flags, and a larger-than-life statue of the Buddha sitting cross-legged, facing other similarly sized statues of his disciples. Plaques of black granite, each etched with words from the Buddha's sermons, in the languages of Buddhist countries worldwide, surrounded the seated figures.

Later in the afternoon, I went to the quiet environs of the Deer Park nearby, and sat beside the Dhamekh Stupa, a staggering Buddhist reliquary nearly 100 feet wide at the foundations and 150 feet tall, its base clad in stones bearing inscriptions. The top half of the stupa was layered in brick. The word "Dhamekh" is derived from Pali, the language of the Buddha's time, and means the "beholding of the dharma," the essence of which was preached by the Buddha in his first sermon at the Deer Park.

I was shielded from the afternoon sun by the stupa; its shade seemed to still the mind. I let myself imagine a time, 2,500 years ago, when a thirty-five-year-old monk preached his radical message of no-self.

● ● ○

Recall the allegory of the man we encountered in the prologue—he who had his body parts replaced by those from a corpse. When the man asked a group of Buddhist monks whether he existed or not, they put the question back to him: who are you? To which the man replied that he wasn't even sure he was a person.

The monks pointed out to him that he had begun realizing that the "I"—his self—is not real. Sure, he had begun doubting whether he existed or not, but the truth is that he had always lacked a self. There was no difference between his old body and his new one, they told him. The feeling "this is my body" was brought about by the aggregation of elements that constituted the body. The man saw the truth of it and was liberated in the Buddhist sense of the word—he was freed of all attachment to things illusory.

I must admit that in 2011, when I visited Sarnath, pondering the Buddhist idea of no-self was intellectually daunting. A lot of what I understood by the term "self" remained an intuition—the kind that we all have about ourselves. What did "no-self" mean when confronted with the intuitive solidity of one's self? When it comes to theories of the self, the target of inquiry is a self that has a perceived unity. There is the unity to everything one is and perceives in any given moment. My sense of being in a body, owning the body, feeling as if I am the agent of my actions, the feeling that everything I perceive is being perceived by me—all of this has a feeling of coherence. There is a single entity that is the subject of experiences, and all the experiences are being had by *me*. This is what philosophers call synchronic unity.

There is also a feeling that this entity endures over time. When you recall your childhood memories, they feel like your memories, and the emotions and perceptions they beget feel like they belong to you. The same is true if you imagine yourself in the future. While we know that we have grown up and changed over time, we have the feeling that underlying all that is the same someone or something, maybe changing, evolving. Philosophers call this diachronic unity.

Both synchronic and diachronic unity were used to argue for the existence of a self very effectively by philosophers of the Nyaya tradition in India (*nyaya* means "logic"), whose earliest texts date back to 200 BCE. Where synchronic unity was concerned, they argued that there must be a self that could collate the various sensations (touch, sight, and hearing, for example) and create a unified perception.

Their position on diachronic unity was more persuasive. They argued that in order for memory to be coherent—meaning that whenever I recall something, it feels like *my* memory—there must be a self. This was a claim that relied on the argument that I cannot recollect your memories and you cannot recall mine. So, if there is no self, then the memory of a past event cannot be recalled as belonging to whoever or whatever is doing the recalling. For memory to function as it does, there must be a self, went their argument. "I'm *not* a big believer in the self, but I think that is the most solid argument that can be made for the self," Georges Dreyfus, a philosopher and scholar of Tibetan Buddhism at Williams College in Williamstown, Massachusetts, told me.

So, broadly speaking, philosophers and neuroscientists fall into two camps: those who claim the self is real and those who say it's not. The big question for them is this: is there an entity that exists that can be called the self, which gives rise to this synchronic and diachronic unity? One hard-nosed way of thinking of the self is to ask whether it can exist independent of all else—as a fundamental part of reality, giving it a unique place in the basic categories, or ontology, of things that make up reality—a self that could not be explained away as being constituted of things with a more basic ontological status. Often, this type of ontologically distinct self is what's being denied by those in the no-self camp—and it's rather easy to do so. No wonder, then, that the pro-self camp thinks it's a straw-man argument.

The experiences of the people in this book, who are suffering from what can be called *maladies of the self*, as well as the neuroscience that explains their experiences, bring us partway to some answers. Aspects of the self that seemingly give us both synchronic and diachronic unity—our narrative, our sense of being agents of our actions and initiators of our thoughts, our sense of ownership of body parts, our sense that we are our emotions, our sense of being located in a volume of space that is our body and possessing a geometric perspective that originates behind our eyes—all of these can be argued as comprising the self-as-object. These properties can be thought of as constructed. The question is whether there is a constructor—or merely the appearance of a constructor.

It's clear that even when these aspects start falling apart, there is still a self-as-subject—what philosophers would call a phenomenal subject—that is consciously experienced. There is still an "I" that is suffering from schizophrenia, is depersonalized, is dealing with autism, is ecstatic, disowns body parts, has out-of-body experiences, loses its narrative, and even denies its own existence. Who or what is that "I"?

A similar approach to getting at the essence of the self was poetically illustrated by Adi Shankara, an eighth-century Indian philosopher and theologian of the Advaita (non-dualist) tradition. His poem, called "Nirvana Shaktam" (the song of liberation in six stanzas), begins with these lines:

> I am not the mind, nor the intellect, nor any entity that
> identifies self with the ears, tongue, nose or the eyes;
> Not even perceived by space, earth, light or the wind.

Each stanza of the poem ends with an answer to the question: who am I? This answer becomes the refrain, building up toward a forceful final stanza. Keeping aside the Advaitan answer for now, the poem's power comes from its assertions of what I am *not*—I am not my mind, my intellect, my body, my senses, my emotions, I am neither virtue nor hatred, I am not my wealth or my relationships, I wasn't even born.

Who am I?

It's this "I" that lies at the heart of the self-versus-no-self debate. What do we make of this self-as-subject, self-as-knower, the experience of subjectivity? Where does that come from? Is there a self or not?

The Buddhist answer—regardless of which of the numerous traditions of Buddhism you pose this question to—is *no*: there is no such thing as a self. If you were to go looking for it (via introspection and meditation), Buddhism argues, you'd arrive at the insight that the self is impermanent, fluctuating, and its perceived unity an appearance.

Within the Western philosophical tradition, eighteenth-century Scottish philosopher David Hume is often quoted for his words:

"When I enter most intimately into what I call myself, I always stumble on some particular perception or other, of heat or cold, light or shade, love or hatred, pain or pleasure. I never can catch myself at any time without a perception, and never can observe any thing but the perception." It's commonly argued that Hume falls into the no-self camp (though philosopher Galen Strawson thinks otherwise, as he argues in his book *The Evident Connexion: Hume on Personal Identity*).

Philosopher Daniel Dennett also belongs to the no-self camp: "Each normal individual of this species makes a *self.* Out of its brain it spins a web of words and deeds and, like other creatures, it doesn't have to know what it's doing; it just does it. . . . Our tales are spun, but for the most part we don't spin them; they spin us." Dennett says the self "is the same kind of thing as a center of gravity [in physics], an abstraction that is, in spite of its abstractness, tightly coupled to the physical world." Any physical system has a center of gravity—but it's not a thing, but a property of the system. There is no one atom or molecule that makes up the center of gravity; nonetheless this mathematical abstraction has real consequences. The self, says Dennett, is the center of *narrative* gravity: a "fiction, posited in order to unify and make sense of an otherwise bafflingly complex collection of actions, utterances, fidgets, complaints, promises, and so forth, that make up a person."

In one sense, the Buddhists, Hume, Dennett, and many others could be categorized as *bundle theorists*: the self, with its perceived unity at any given moment and over time, is "entirely fabricated from the bundle of discrete mental phenomena."

Thomas Metzinger is also a no-self theorist. We have already encountered his position. He posits that an ongoing biological process, rooted deeply in the body, gives rise to a representation—a self-model—of the organism in the brain. The contents of this dynamic self-model include everything from the body and its emotional state to sensations and thoughts. The content of your self-model makes up everything you can consciously experience about yourself. Crucially, the conscious self-model is transparent, which means that we don't experience the self as a representation, even if intellectually we believe (and can maybe someday prove) it to be so. "It is a very robust, reality-rendering mechanism," Metzinger told me. To him, the experience of being a phenomenal self—the self as subjectively experienced—comes from being conscious of the interactions between the self-model and the world-model. Metzinger's ideas do away with the self as being an entity or a thing that would persist outside of the living brain. But the exact neural processes that could give rise to the subjectivity as modeled by Metzinger are yet unclear.

Another perspective comes from Antonio Damasio. Recall his framework of the protoself, the core self, and the autobiographical self: these components comprise the self-as-object. To this he adds a self-as-knower or self-as-subject—some neural process in the brain that gives us the experience of the self as something that knows itself, and gives the mind subjectivity: "When the brain manages to introduce a knower in the mind, subjectivity follows." To put it simply, the self-as-knower makes us conscious. Philosopher John Searle, in his critique of Damasio's *Self Comes to Mind*, argued that this is circular reasoning: "The self is introduced to explain consciousness, but if it is to explain consciousness we cannot assume that the self is already conscious."

The criticism highlights the challenges neuroscientists and philosophers face in explaining the subjectivity of self-consciousness. It's no wonder, then, that some philosophers simply attribute subjectivity to consciousness itself (keeping aside the hard problem for now). They think that underlying all our experiences are conscious states that have the property of self-awareness. Note, they are not saying that there is a subject, or someone, who is experiencing. Rather, that consciousness has the property of subjectivity. Consciousness is reflexive, in philosophical jargon. "Reflexivity is something automatic, pervasive, passive, something characterizing consciousness from the start," philosopher Dan Zahavi of the University of Copenhagen told me.

So, in this way of thinking, the brain must somehow take these moments of self-awareness, or moments of reflexive consciousness, and construct a self that appears unitary and solid. But neuroscience is far, far away from explaining how reflexivity of consciousness may arise. However, if you take this property of consciousness as given, some no-self theorists would say that there is no self, just moments of reflexive consciousness.

Jonardon Ganeri, a philosopher of mind at New York University, thinks that if you accept consciousness as being inherently reflexive, that's as good as saying there is a self. "You wonder what this denial of self is meant to be doing," he told me. "Why not just say that what selfhood consists [of] is the reflexivity of consciousness. That seems like a pretty good account of self to me." But Ganeri acknowledges that even if consciousness is reflexive, it does not disprove or negate the existence of a self that stands on its own, alongside reflexive consciousness.

Zahavi argues for such a self: a minimal self—which, he says, provides the mental structure to make an experience seem like *mine*, to give it a first-person perspective. Such a minimal self has to transcend or linger for longer than any given moment of subjectivity—such that many such moments can be experienced as belonging to the same subject.

Take the case of people with schizophrenia, who lose sense of ownership of their own thoughts at times. "Something minimal has to be preserved to even make sense out of the disorder," said Zahavi. In all the experiences of people with perturbations of the self that we encountered, the sense that something minimal survives is hard to dismiss—whatever the experience, whether of depersonalization or being out-of-body, the associated sense of *mineness* of experience remained. "It's really hard to think of a scenario that could potentially give us a case of experience completely devoid of any kind of minimal ownership," Zahavi told me. "How should it even be reported in the first person?"

It's in order to answer that latter question that Zahavi posits his minimal self. But that then brings up other questions. Explaining the subjective character of the minimal self is no easier than explaining how consciousness arises (in fact, Zahavi rejects the idea that the "self is something separate from and independent of consciousness"; he told me, "We cannot understand and do justice to consciousness without operating with a minimal notion of self").

So, how does this minimal self, which provides synchronic unity, get extended to form one's entire selfhood, with diachronic unity? Zahavi thinks that we need something in between these two extremes of the minimal self and the full-blown extended, narrative self: a form of interpersonal self that is built up from the minimal self during early childhood as the infant interacts with its mother and others, when there isn't a full-blown narrative self yet but nonetheless the infant is developing a self in relation to others.

At another end of the spectrum of theorists are the Indian Advaita (non-dualist) thinkers. They argue that there is an underlying, un-individuated consciousness that is the subject of all experiences—not just of yours or mine, but of all experiences—a consciousness that is witness to everything. An impersonal experiencer. It's the denouement of Adi Shankara's six-stanza poem.

So, while Advaita philosophers concur with the no-self idea, maintaining that the individual self is not real, they eventually diverge from their Buddhist brethren. The Buddhist version of the bundle theory says that our "mistake lies in taking there to be one thing when there are really only the many," or mistakenly perceiving the bundle as real when there are only many interacting psycho-physical elements. The Advaita philosophers argue that "our mistake lies in taking there to be a many when strictly speaking there is just the one"—the consciousness that experiences everything.

It's hard to escape the feeling that neuroscientists and philosophers (both past and present), in their arguments over whether there is a self or not, are converging—or, dare I say, splitting hairs. There's very little they are at odds about. Descartes's dualism is passé. No one is arguing for a self that has an independent ontological reality, something that could exist even after the brain and body are gone. No one's arguing either for a single privileged place in the brain as the sole custodian of the self. Yes, there are some brain regions that are more important than others for our sense of self—such as the insular cortex, the temporoparietal junction, and the medial prefrontal cortex—but none that can be said to be the singular domain of the self. There's also little argument that our narrative self is a fiction—a story without a storyteller. In fact, anything that can constitute the self-as-object—including the sense of body ownership—can be argued as being constructed, sans a constructor. Instead of Cartesian dualism, which relegated the body to the status of a mere vessel, we now have a picture of the sense of self as the outcome of neural processes that are tightly integrated with the body—processes that combine brain, body, mind, and even culture to make us who we are. What remains to be satisfactorily explained is the self-as-subject or self-as-knower, and that's where the differences arise. The subjectivity of experience: just how does it come to be? Whether that subjectivity is due to some neural process that Zahavi would call the minimal self, or is due to an inherent reflexivity of consciousness, or is something that appears to be so because of the interaction of psycho-physical elements (à la bundle theorists or Dennett or Metzinger)—that's where the mystery of the self now resides. Most likely, solving the mystery will require making sense of consciousness itself.

● ● ○

Besides the intellectual and philosophical wrangling, there is human suffering. From the perspective of the experiences of people we met in this book, understanding the nature of the self is crucial. If, as the Buddhists claim, it's our delusional attachment to a seemingly solid self that causes suffering, then realizing its true nature could ease suffering (and as we have seen, the various aspects of the self that make up the self-as-object really appear because of brain dynamics, something that one can get dissociated from). Ganeri pointed out to me that the Buddhist take on the suffering of individuals with maladies of the self would be to say that we set the benchmark for the self too high in the first place. So, the disturbances appear as deficits, and everything else—the coping mechanisms, the treatments, and the therapies—ensues from that understanding. But what if the disturbances were seen as the outcome not of deficits of self but of an obsessive attachment to the idea of a self? Letting go could have therapeutic benefits.

I am reminded of my discussion with Jeff Abugel, who has endured periods of depersonalization since his late teens, and whom we briefly encountered in chapter 5. Abugel told us how his whole life had become an exercise in trying to figure out—on a moment-by-moment basis—what was wrong with him, why he felt so estranged from himself. Medication helped, up to a point. "The medication for me is just to reduce the way of fragmented thinking, the way of thinking that was very uncomfortable, simply because I didn't feel integrated, I felt detached, I felt broken up," he told me. "The medicine has helped to integrate my sense of self, but it didn't restore the sense of self that I had when I was eighteen." To make sense of his disintegrated state, he turned to writings of philosophers who were suggesting that the dissipation of the ego and a breakdown of its perceived unity brought about a new state of being. "I think in very ancient cultures there are parallels that are very easily drawn between the depersonalization as we experience it and how these other people either sought it, or looked for it, or felt it, and then tried to make sense of it," said Abugel. He has, in some sense, let go of aspects of his former self, or at least stopped striving to get it back. "From the patient's standpoint, you have really got two choices: you can keep trying every kind of medicine and therapy until you regain the sense of self and ego that you had before it all started, or you can say, 'OK, I got fifty percent of it back, let me see where the other fifty percent goes. Let me see what it's all about.'"

It's been a somewhat beneficial journey. "To me the big question is: Do you view [depersonalization] as a disorder, or do you view it as a different state of mind? Do you view it, for a lack of a better word, as some kind of beginning on a road to some kind of awakening?" Abugel told me. "In time, I have come to view it as simply a change in perception. It's changed my view of the world, seeing as how fleeting and small it actually is, compared to all of existence."

Of course, being able to do what Jeff is doing presupposes some amount of cognitive ability. Someone who is suffering from severe schizophrenia or autism, or is in the throes of Cotard's syndrome, unfortunately cannot escape his or her phenomenal self: all the talk of the self being a construction sans a constructor will have no effect. Their suffering is real. It's also unrealistic to expect a person with Alzheimer's to cope with the loss of the narrative self by focusing on the fact that there is no narrator to begin with.

But some—those with mild forms of schizophrenia, depersonalization, or BIID, perhaps—may find therapeutic succor by gaining insights into the nature of the self. However, it's not just those with maladies of the self who could benefit from such insights.

● ● ○

There must have been a time in our evolutionary past when the first glimmers of the self-as-knower appeared. It must have been a momentous biological event. And it gave our ancestors a survival advantage. To be aware of one's own body, to be able to direct one's attention to it, must have been an evolutionary leg up. But this self-process—a complex interaction of the activity of various brain regions—was still meant to control one's body. As we evolved further, we developed the various forms of long-term memory, a narrative self; we could learn from our mistakes, and we could plot and plan our future. Thoughts about our past and future selves were added to the mix that was the self-as-object. We went from creatures that lived in the here and now to creatures that inhabited a mental time line. However, no matter how ethereal our thoughts, the feedback as to whether those thoughts were good or bad for one's conception of oneself was still mediated by the body. It could be a sense of elation, or a sinking feeling in the pit of one's stomach, or myriad variations on the theme, from ecstasy to depression. These feelings and emotions were meant to make us act, to move us toward pleasure and joy, away from pain and sadness. Except that where we once felt these emotions because we moved toward a source of food in the forest or fled a predator, we could now feel them because of the contents of our thoughts, even when they had no direct bearing on survival. This has, of course, made us the species we are, with society and culture and art and technology and all that's beautiful about being human. It's also turned us into a species that cannot stop from wanting more. For most of us, imagining having more makes us feel good and safe, imagining having less has the opposite effect, and we act on these feelings. We are now beholden to the survival of the conceptual self and not just the bodily self—and this imagined self has no limits to its grandiosity. So, while the vexing nature of the self has given us ascetics and monastics, who have enquired into its nature with all their being, this runaway process has also given us narcissism and overindulgence. It would not be a stretch to say that many of society's ills can be attributed to an unbridled conceptual self, which wants too much or fights to preserve reified identities: the ideological stubbornness of religions; the growing disparity between the wealthy and the poor; the hegemonies of powerful, militarized nations over smaller ones; or the continued plunder of natural resources.

Coming to terms with the self's mostly fictitious nature (the unresolved issue of subjectivity notwithstanding) may help rein ourselves in. It's unclear, though, whether mere intellectual understanding will do. In fact, the Buddhists developed the notion of no-self not as an intellectual argument but to give philosophical heft to an experience that arises out of meditation. "No question that no-self is a very important idea," said Georges Dreyfus, who is in the no-self camp. "But it's trying to capture an experience which happens to people mostly through meditation. It's an experience which has profound transformative effects, in terms of diminishing self-centeredness, making oneself more open to others, and so on."

In Buddhism and Advaita, the no-self idea arises out of concern for the suffering of people. Mistakenly identifying with *me* and *mine* is at the root of suffering, they say. Realizing this and losing one's attachment to the self is liberation, the end of suffering. "The core Buddhist thought is that cognitive attachments to self are itself a kind of pathology, a kind of a source of dysfunction," said Jonardon Ganeri.

The malady *is* the self.

ACKNOWLEDGMENTS

This book begins with an ancient Buddhist allegory from about 200 CE—sincere thanks to Jonardon Ganeri for alerting me to the allegory and for letting me use the English translation from his book *The Self.*

My book would have been impossible to write without the extraordinary kindness and openness of the many people who shared their stories with me. Some of you I have to thank using pseudonyms, but you know who you are.

Thanks to Michaele and Allan for inviting me into their home and spending time talking about Allan's Alzheimer's—Allan, sadly, passed away soon after I met him. To Clare for sharing her father's story and letting me visit him at his long-term care home. Unfortunately, Clare's father is in the late stages of Alzheimer's, and I can only wonder about what he understood when Clare introduced us. My thanks to him.

Thanks to Patrick and David—two men who let me into their lives and put their trust in me, despite the risks of do so. I'm extremely grateful to both of them for letting me join them on their journey to see Dr. Lee, who let me witness the process of David getting his leg amputated (short of being in the OR). The story first appeared in *MATTER* magazine in October 2012; thanks to Jim Giles, Bobbie Johnson, and Roger Hodge for their part in making it happen.

Thanks to Laurie and Sophie for sharing their very personal—often harrowing—experiences of schizophrenia; their keen insight into their own condition proved invaluable to someone like me, who was trying to make sense of another's altered sense of reality. And to Laurie's husband, Peter, for providing a partner's perspective.

I'm grateful to Nicholas and his fiancée, Jasmine, for helping me understand Nicholas's depersonalization, and to Nick's foster mother, Tammy, and his physician for sharing their side of things. Thanks to Jeff Abugel for his insights into his own condition and for connecting me to Nicholas. And to Sarah for taking the time to sit down and talk about her transient, but nonetheless scary, encounter with depersonalization, and to Ellen Petry Leanse for the introduction.

James Fahey, an articulate advocate for those with Asperger's syndrome, helped me understand what it means to be an adult with his condition and yet live without being constrained by psychiatric definitions and socially imposed boundaries. My sincere thanks to him. And to my friends Susan and Roy, and their son Alex, for letting me into their lives in so many ways and for being open about Alex's and their struggles.

My heartfelt thanks to my cousin Shobha and her husband, Ashok, for talking about Ashwin so soon after he passed away—it must have been hard, as parents—and for letting me write about his doppelgänger experience. Thanks to Chris and Sonia for taking an emotional trip back in time; talking about Chris's brother, who died too young, wasn't easy for them. To Thomas Metzinger for being open about his out-of-body experiences and for patiently explaining his philosophy of the self, over and over.

I'm grateful to Zachary Ernst for an insightful take on his ecstatic seizures; his philosopher's acumen helped me make sense of a seemingly mystical experience. To Catherine and Albéric, mother and son, for welcoming me to Romont, where, amid staggeringly beautiful Swiss alpine scenery, Albéric spoke in French about his ecstatic epilepsy and Catherine translated.

Thanks also to many of the aforementioned people for reading and checking what I wrote about them.

My gratitude to the many scientists, doctors, and philosophers whom I quote in the book (and some whom I don't): they gave generously of their time and energy and shared their expertise—in person and by phone or email; many of them read parts of the book and made valuable suggestions and corrections. I'm grateful to (in the order of chapters): Adam Zeman, David Cohen, Steven Laureys, William de Carvalho, Athena Demertzi, Lionel Naccache, Shaun Gallagher, Pia Kontos, Robin Morris, William Jagust, Suzanne Corkin, Bruce Miller, Giovanna Zamboni, Paul McGeoch, Michael First, Judith Ford, Ralph Hoffman, Gottfried Vosgerau, Martin Voss, Nick Medford, Hugo Critchley, Sanjeev Jain, R. Raguram, Alison Gopnik, Uta Frith, Elizabeth Torres, Francesca Happé, Peter Hobson, Philippe Rochat, Jakob Hohwy, Peter Enticott, Olaf Blanke, Lukas Heydrich, Bigna Lenggenhager, Henrik Ehrsson, Arvid Guterstam, Manos Tsakiris, Thomas Grunwald (who arranged for me to witness a neurosurgery), Fabrice Bartolomei, Bud Craig, Antoine Bechara, Jonardon Ganeri, Dan Zahavi, Georges Dreyfus, and Geshe Ngawang Samten.

And I am especially grateful to Peter Brugger, Louis Sass, Anil Seth, Thomas Metzinger, and Fabienne Picard, all of whom went far beyond what could be reasonably expected, endured endless emails and phone calls, hosted my visits to their offices and labs and even homes, and vetted parts of the book, all with boundless generosity.

Thanks to my friends: Caroline Sidi for help with all things French; Srinath Perur for reading and commenting on the whole book; Rajesh Kasturirangan and Vikramajit Ram for invaluable inputs; Venu Narayan for being a sounding board from start to finish. C. S. Aravinda translated "Nirvana Shaktam" from Sanskrit to English—my thanks to him.

Needless to say, any errors that remain are my sole responsibility.

The book proposal took shape with the help of my agent, Peter Tallack, at the Science Factory. Thanks, as always, Peter. Thanks also to my editor, Stephen Morrow, for being an astute and attentive listener, for seeing what I saw in these stories, and for gently guiding the book to its finish.

My friends and colleagues at *New Scientist*—thanks for doing what you do. I learned my craft there, and I continue to do so.

Also, a warm thanks to my friends who made me feel at home in their homes during my peregrinations as I researched this book: Caroline Sidi, Alok Jha, Banu and Ramesh, Vijay and Hema, Maithili and Prasad, Rao and Kinkini, Anjali and Kiran, and Suruchi and Biraj.

I got through the final three months of frenetic writing because my mother and father took care of everything else—from the regular cups of coffee to nourishing vegetable stews. So, finally, to my parents and sisters, brothers-in-law, and niece and nephews, a big thank-you for being family, for all your support and love.

NOTES

ix **"It seems outlandish":** Thomas Nagel, *The View from Nowhere* (Oxford: Oxford University Press, 1986), 55.

PROLOGUE

1 **An allegory about a man:** Parable adapted with permission from Jonardon Ganeri. The English translation appears in Jonardon Ganeri, *The Self: Naturalism, Consciousness and the First-Person Stance* (Oxford: Oxford University Press, 2012), 115.

CHAPTER 1: THE LIVING DEAD

3 **"Men ought to know":** Quoted in Adam Zeman, "What in the World Is Consciousness?," *Progress in Brain Research* 150 (2005): 1–10.

3 **"If I try to seize":** Albert Camus, *The Myth of Sisyphus and Other Essays* (New York: Vintage, 1991), 19.

6 **the façade, as the architect intended:** Michel Delon, ed., *Encyclopedia of the Enlightenment* (London, New York: Routledge, 2013), 258.

7 **Cotard had been devotedly:** J. Pearn and C. Gardner-Thorpe, "Jules

Cotard (1840–1889): His Life and the Unique Syndrome Which Bears His Name," *Neurology* 58 (May 2002): 1400–03.

7 **the case of a forty-three-year-old woman:** G. E. Berrios and R. Luque, "Cotard's Delusion or Syndrome?: A Conceptual History," *Comprehensive Psychiatry* 36, no. 3 (May/June, 1995): 218–23.

8 **"clear and distinct intellectual":** "René Descartes," *Stanford Encyclopedia of Philosophy*, http://plato.stanford.edu/entries/descartes.

8 **"one cannot be wrong":** Thomas Metzinger, "Why Are Identity Disorders Interesting for Philosophers?," in *Philosophy and Psychiatry*, Thomas Schramme and Johannes Thome, eds. (Berlin: Walter de Gruyter GmBH & Co, 2004), 311–25.

8 **"Patients may explicitly":** Ibid.

9 **"Who is the I":** Gordon Allport, quoted in Stanley B. Klein and Cynthia E. Gangi, "The Multiplicity of Self: Neuropsychological Evidence and Its Implications for the Self as a Construct in Psychological Research," *Annals of the New York Academy of Sciences* 1191 (March 2010): 1–15.

9 **We instinctively and intimately:** Anil Ananthaswamy, "Am I the Same Person I Was Yesterday?" *New Scientist*, July 23, 2011.

9 **"Know thyself":** Pausanias, *Description of Greece*, http://www.perseus.tufts.edu/hopper/text?doc=Perseus:text:1999.01.0160:book=10:chapter=24.

9 **"By whom commanded and":** Swami Paramananda, *The Upanishads* (The Floating Press, 2011), 69.

9 **"If no one asks":** Saint Augustine quoted in Klein and Gangi, "The Multiplicity of Self."

10 **fifteen-year-old May:** David Cohen et al., "Cotard's Syndrome in a 15-Year-Old Girl," *Acta Psychiatrica Scandinavica* 95 (February 1997): 164–65.

11 **often an effective treatment:** For an eloquent defense of this seemingly barbaric procedure, see Sherwin Nuland's TED talk titled "How Electroshock Therapy Changed Me," http://www.ted.com/talks/sherwin_nuland_on_electroshock_therapy?language=en.

12 **a fifty-five-year-old man:** David Cohen and Angèle Consoli, "Production of Supernatural Beliefs during Cotard's Syndrome, a Rare Psychotic Depression," *Behavioral and Brain Sciences* 29, no. 5 (October 2006): 468–70.

12 **"God's punishment for sins":** Ibid.

14 **"a facial expression involving":** Edward Shorter, "Darwin's Contribution to Psychiatry," *The British Journal of Psychiatry* 195, no. 6 (2009): 473–74.

14 **"grief muscles":** Ibid.

14 **"melancholic omega":** Ibid.

14 **the French philosopher Louis:** "Louis Althusser," *Stanford Encyclopedia of Philosophy,* http://plato.stanford.edu/entries/althusser.

17 **the frontoparietal network:** Audrey Vanhaudenhuyse et al., "Two Distinct Neuronal Networks Mediate the Awareness of Environment and of Self," *Journal of Cognitive Neuroscience* 23, no. 3 (March 2011): 570–78.

18 **the dubious field of phrenology:** "Franz Joseph Gall," *Encyclopedia Britannica,* http://www.britannica.com/EBchecked/topic/224182/Franz-Joseph-Gall.

19 **the extent of the lowered metabolism:** Vanessa Charland-Verville et al., "Brain Dead Yet Mind Alive: A Positron Emission Tomography Case Study of Brain Metabolism in Cotard's Syndrome," *Cortex* 49 (2013): 1997–999.

19 **some regions that are:** In Graham's case, these were the dorso-lateral prefrontal regions.

19 **a sixty-five-year-old woman with dementia:** Seshadri Sekhar Chatterjee and Sayantanava Mitra, "'I Do Not Exist': Cotard Syndrome in Insular Cortex Atrophy," *Biological Psychiatry* (November 2014).

21 **the immunity principle:** Shaun Gallagher, "Philosophical Conceptions of the Self: Implications for Cognitive Science," *Trends in Cognitive Sciences* 4, no. 1 (January 2000): 14–21.

22 **at least three such facets:** William James, *The Principles of Psychology,* https://ebooks.adelaide.edu.au/j/james/william/principles/chapter10.html.

22 **"a man has as many":** Ibid.

22 **"a man's inner or subjective":** Ibid.

22 **the *mineness*:** Thomas Metzinger, *Being No One: The Self-Model Theory of Subjectivity* (Cambridge, MA: MIT Press, 2003), 267.

23 **"Something has happened to me":** Quoted in Sue E. Estroff, "Self, Identity, and Subjective Experiences of Schizophrenia: In Search of the Subject," *Schizophrenia Bulletin* 15, no. 2 (1989): 189.

24 **Emerson was curiously indifferent:** Elizabeth Arledge, *The Forgetting: A Portrait of Alzheimer's*, PBS, 2004, http://www.pbs.org/theforgetting/experience/first_person.html.

24 **he could still conduct:** Ibid.

CHAPTER 2: THE UNMAKING OF YOUR STORY

27 **"Memory, connecting inconceivable mystery":** Ralph Waldo Emerson, *The Later Lectures of Ralph Waldo Emerson, 1843–1871,* vol .2, Ronald A. Bosco and Joel Myerson, eds. (Athens and London: University of Georgia Press, 2010), 102.

29 **A handwritten note:** Konrad Maurer et al. "Auguste D and Alzheimer's Disease," *Lancet* 349, no. 9064 (May 1997): 1546–549.

29 **"She sits on the bed":** Ibid.

30 **"On what street":** Ibid.

30 **"sampled thin slices":** "History Module: Dr. Alois Alzheimer's First Cases," http://thebrain.mcgill.ca/flash/capsules/histoire_jaune03.html.

30 **"Alzheimer put down":** David Shenk, "The Memory Hole," *New York Times*, November 3, 2006.

30 **"progressive cognitive impairment":** Maurer et al., "Auguste D."

30 **"A Characteristic Serious Disease":** Ibid.

30 **"In the centre of":** Ibid.

31 **"miliary foci":** Ibid.

31 **"The clinical interpretation":** Ibid.

34 **"Alzheimer's disease robs you":** Elizabeth Arledge, *The Forgetting*, 1.40 sec. Italics mine.

35 **Clare, a sixty-year-old woman:** Some identifying details about Clare and her father have been changed upon Clare's request, including her name.

35 **Consider the phrases used:** For a critique, see Pia C. Kontos, "Embodied Selfhood in Alzheimer's Disease: Rethinking Person-Centred Care," *Dementia* 4, no. 4: 553–70.

36 **"Individuals construct private":** Donald E. Polkinghorne, "Narrative and Self-Concept," *Journal of Narrative and Life History* 1, nos. 2–3 (1991): 135–53.

36 **"the self is ultimately":** Joel W. Krueger, "The Who and the How of Experience," in *Self, No Self? Perspectives from Analytical, Phenomenological, & Indian Traditions*, Mark Siderits et al., eds. (Oxford: Oxford University Press, 2011), 37.

37 **"It is by no means":** Dan Zahavi, "Self and Other: The Limits of Narrative Understanding," *Royal Institute of Philosophy Supplement* 60 (May 2007), 179–202.

41 **the causation was unclear:** Suzanne Corkin, "Lasting Consequences of Bilateral Medial Temporal Lobectomy: Clinical Course and Experimental Findings in H.M.," *Seminars in Neurology* 4, no. 2 (June 1984): 249–59.

41 **"could no longer recognize":** William Beecher Scoville and Brenda Milner, "Loss of Recent Memory after Bilateral Hippocampal Lesions," *Journal of Neurology, Neurosurgery & Psychiatry* 20 (1957): 11–21.

41 **"This was performed on April 26":** Ibid.

42 **"A striking feature of H.M.":** Corkin, "Lasting Consequences."

44 **the back half of:** Suzanne Corkin et al., "H. M.'s Medial Temporal Lobe Lesion: Findings from Magnetic Resonance Imaging," *The Journal of Neuroscience* 17, no. 10 (May 1997): 3964–979.

44 **"It is the most heavily":** Gary W. Van Hoesen et al., "Entorhinal Cortex Pathology in Alzheimer's Disease," *Hippocampus* 1, no. 1 (January 1991): 1–8.

44 **"unforgettable amnesiac":** Benedict Carey, "H. M., an Unforgettable Amnesiac, Dies at 82," *New York Times*, December 4, 2008.

45 **the same brain networks that:** Daniel L. Schacter et al., "The Future of Memory: Remembering, Imagining, and the Brain," *Neuron* 76, no. 4 (November 2012): 677–94.

45 **"I want to draw attention":** Quoted in Errol Morris, "The Anosognosic's Dilemma: Something's Wrong but You'll Never Know What It Is (Part 2)," *New York Times*, June 21, 2010, http://opinionator.blogs.nytimes.com/2010/06/21/the-anosognosics-dilemma-somethings-wrong-but-youll-never-know-what-it-is-part-2.

46 **"If she was asked":** Quoted in Ibid.

47 **the medial prefrontal cortex:** Giovanna Zamboni et al., "Neuroanatomy of Impaired Self-Awareness in Alzheimer's Disease and Mild Cognitive Impairment," *Cortex* 49, no. 3 (March 2013): 668–78.

48 **"What is your favorite":** Suzanne Corkin, "What's New with the Amnesic Patient H.M.?," *Nature Reviews Neuroscience* 3 (February 2002), 153–60.

52 **reminiscence bump:** Clare J. Rathbone et al., "Self-Centered Memories: The Reminiscence Bump and the Self," *Memory & Cognition* 36, no. 8 (2008): 1403–414.

53 **"constrains what the self":** Martin A. Conway, "Memory and the Self," *Journal of Memory and Language* 53 (2005): 594–628.

54 **memories of those experiences:** Ibid.

56 **"the idea that bodily habits":** Pia C. Kontos, "Alzheimer Expressions or Expressions Despite Alzheimer's?: Philosophical Reflections on Selfhood and Embodiment," *Occasion: Interdisciplinary Studies in the Humanities* 4 (May 2012): 1–12.

56 **"Nothing human is altogether":** Quoted in *The Embodied Self: Dimensions, Coherence and Disorders,* Thomas Fuchs et al., eds. (Stuttgart: Schattauer GmbH, 2010), v.

56 **"Knowledge of typing":** Kontos, "Embodied Selfhood."

57 **"Habitus comprises dispositions":** Pia C. Kontos, "Ethnographic Reflections on Selfhood, Embodiment and Alzheimer's Disease," *Ageing & Society* 24, no. 6 (2004): 829–49.

57 **"a way of being":** Pia C. Kontos, "Habitus: An Incomplete Account of Human Agency," *The American Journal of Semiotics* 22, no. 1/4 (2006): 67–83.

CHAPTER 3: THE MAN WHO DIDN'T WANT HIS LEG

63 **"The leg suddenly assumed":** Oliver Sacks, *A Leg to Stand On* (New York: Touchstone, 1998), 53.

63 **"Theoretically you could":** V. S. Ramachandran in Christopher Rawlence, *Phantoms in the Brain,* 2000, https://www.youtube.com/watch?feature=player_embedded&list=PL361F982E5B7C1550&v=PpEpjJgGDI#t=138.

66 **They have also suggested Xenomelia:** Paul D. McGeoch et al., "Xenomelia: A New Right Parietal Lobe Syndrome," *Journal of Neurology, Neurosurgery & Psychiatry* 82 (2011): 1314–319.

67 **"an invincible obstacle":** Leonie Maria Hilti and Peter Brugger, "Incar-

nation and Animation: Physical Versus Representational Deficits of Body Integrity," *Experimental Brain Research* 204, no. 3 (2010): 315–26.

67 **The first modern account:** John Money et al., "Apotemnophilia: Two Cases of Self-Demand Amputation as a Paraphilia," *Journal of Sex Research* 13, no. 2 (May 1977): 115–25.

67 **homosexuality was also labeled:** David L. Rowland and Luca Incrocci, eds., *Handbook of Sexual and Gender Identity Disorders* (Hoboken, NJ: John Wiley & Sons, 2008), 496.

68 **The patient died of gangrene:** "Complete Obsession," transcript, BBC, February 17, 2000, http://www.bbc.co.uk/science/horizon/1999/obsession_script.shtml.

68 **A Scottish surgeon named Robert Smith:** Ibid.

68 **First embarked on a survey:** Michael B. First, "Desire for Amputation of a Limb: Paraphilia, Psychosis, or a New Type of Identity Disorder," *Psychological Medicine* 35, no. 6 (June 2005): 919–28.

69 **a meeting in New York:** "Meetings," BIID.ORG, undated, http://www.biid.org/meetings.html.

73 **"It seems like my body":** "Complete Obsession."

73 **"I have become convinced":** Ibid.

74 **cognitive scientists at Carnegie Mellon University:** Matthew Botvinick and Jonathan Cohen, "Rubber Hands 'Feel' Touch That Eyes See," *Nature* 391 (February 19, 1998): 756.

75 **coined the phrase "phantom limb":** See V. S. Ramachandran and William Hirstein, "The Perception of Phantom Limbs: The D. O. Hebb Lecture," *Brain* 121 (1998): 1603–630.

76 **body parts that were absent:** Peter Brugger et al., "Beyond Re-membering: Phantom Sensations of Congenitally Absent Limbs," *Proceedings of the National Academy of Sciences* 97, no. 11 (May 2000): 6167–172.

77 **this area is thinner:** L. M. Hilti et al., "The Desire for Healthy Limb Amputation: Structural Brain Correlates and Clinical Features of Xenomelia," *Brain* 136, no. 1 (January 2013): 318–29.

77 **the right SPL showed reduced:** McGeoch et al., "Xenomelia."

77 **This network, they suggest:** Lorimer G. Moseley et al., "Bodily Illusions in Health and Disease: Physiological and Clinical Perspectives and the Concept of a Cortical 'Body Matrix,'" *Neuroscience and Biobehavioral Reviews* 36, no. 1 (2012): 34–46.

78 **"regulate its interaction":** Thomas Metzinger, "The Subjectivity of Subjective Experience: A Representationalist Analysis of the First-Person Perspective," *Networks* 3–4 (2004): 33–64.

79 **"any regulator":** Roger C. Conant and Ross W. Ashby, "Every Good Regulator of a System Must Be a Model of That System," *International Journal of Systems Science* 1, no. 2 (1970): 89–97.

79 **the property of *mineness*:** Thomas Metzinger, *Being No One: The Self-Model Theory of Subjectivity* (Cambridge, MA: MIT Press, 2003), 267.

80 **a simple and elegant experiment:** David Brang et al., "Apotemnophilia: A Neurological Disorder," *NeuroReport* 19, no. 13 (August 2008): 1305–306.

80 **"desired line of amputation":** Ibid.

81 **their brains were prioritizing:** Atsushi Aoyama et al., "Impaired Spatial-Temporal Integration of Touch in Xenomelia (Body Integrity Identity Disorder)," *Spatial Cognition & Computation* 12, nos. 2–3 (2012): 96–110.

82 **"absolute, utter lunacy":** Randy Dotinga, "Out on a Limb," *Salon*, August 29, 2000, http://www.salon.com/2000/08/29/amputation.

86 **Working swiftly, he bandaged:** Minor details about Dr. Lee's presurgery preparation have been changed to protect him and his medical staff.

87 **The hospital itself:** Some details about David, Patrick, Dr. Lee, the hospital, and its environs were changed to protect the identities of those concerned.

CHAPTER 4: TELL ME I'M HERE

93 **Tell Me I'm Here:** Title of chapter comes from a book of the same name. Anne Deveson (New York: Penguin, 1992).

93 **"What gives me the right":** Quoted in Louis A. Sass, *Madness and Modernism: Insanity in the Light of Modern Art, Literature, and Thought* (Cambridge, MA: Harvard University Press, 1994), 216.

93 **"For any true grasp of delusion":** Karl Jaspers, *General Psychopathology* (Manchester: Manchester University Press, 1963), 97.

93 **I met Laurie and her husband, Peter:** Some sensitive details, including their names, have been changed.

95 **"Sorry, I'm Sophie":** Some identifying details, including her name, have been changed.

98 **his 1992 book:** Sass, *Madness and Modernism.*

104 **"the study of 'lived experience'":** Louis A. Sass and Josef Parnas, "Schizophrenia, Consciousness, and the Self," *Schizophrenia Bulletin* 29, no. 3 (2003): 427–44.

105 **"This experience of one's *own*":** Louis A. Sass, "Self-Disturbance and Schizophrenia: Structure, Specificity, Pathogenesis (Current Issues, New Directions)," *Schizophrenia Research* 152, no. 1 (January 2014): 5–11.

109 **Charles Bell and Johannes Purkinje:** Bruce Bridgeman, "Efference Copy and Its Limitations," *Computers in Biology and Medicine* 37, no. 7 (July 2007): 924–29.

109 **in 1950, Erich von Holst and Horst Mittelstaedt:** Erich von Holst and Horst Mittelstaedt, "Das Reafferenzprinzip," *Die Naturwissenschaften* 37, no. 20 (October 1950): 464–76. Translated as: "The Principle of Reafference: Interactions between the Central Nervous System and the Peripheral Organs," in *Perceptual Processing: Stimulus Equivalence and Pattern Recognition*, P. C. Dodwell, ed. (New York: Appleton-Century-Crofts, 1971), 41–72.

109 **"Eristalis has a slender":** Ibid.

109 **"Its small size":** Roger Sperry, "Neural Basis of the Spontaneous Optokinetic Response Produced by Visual Inversion," *Journal of Comparative and Physiological Psychology* 43, no. 6 (December 1950): 482–89.

111 **"The subjective experience":** Irwin Feinberg, "Efference Copy and Corollary Discharge: Implications for Thinking and Its Disorders," *Schizophrenia Bulletin* 4, no. 4 (1978): 636–40.

111 **"Thus, if corollary discharge":** Ibid.

113 **the cricket tunes in:** James F. A. Poulet and Berthold Hedwig, "The Cellular Basis of a Corollary Discharge," *Science* 311 (January 27, 2006): 518–22.

114 **It's near impossible to tickle yourself:** Sarah-Jayne Blakemore et al., "Why Can't You Tickle Yourself?," *NeuroReport* 11, no. 11 (August 2000): R11–16.

114 **people experiencing auditory hallucinations:** Sarah-Jayne Blakemore et al., "The Perception of Self-Produced Sensory Stimuli in Patients with Auditory Hallucinations and Passivity Experiences: Evidence for a

Breakdown in Self-Monitoring," *Psychological Medicine* 30, no. 5 (September 2000): 1131–139.

115 **possible disruption of the copy mechanism:** Daniel H. Mathalon and Judith M. Ford, "Corollary Discharge Dysfunction in Schizophrenia: Evidence for an Elemental Deficit," *Clinical EEG and Neuroscience* 39, no. 2 (2008): 82–86.

116 **one's sense of agency should be subdivided:** Matthis Synofzik et al., "Beyond the Comparator Model: A Multifactorial Two-Step Account of Agency," *Consciousness and Cognition* 17, no. 1 (March 2008): 219–39.

116 **they tend to rely more:** Matthis Synofzik et al., "Misattributions of Agency in Schizophrenia Are Based on Imprecise Predictions about the Sensory Consequences of One's Actions," *Brain* 133 (January 2010): 262–71.

117 **"support the notion of":** Ibid.

118 **In one harrowing section:** Deveson, *Tell Me I'm Here,* 132.

120 **there is hyperconnectivity:** Ralph E. Hoffman and Michelle Hampson, "Functional Connectivity Studies of Patients with Auditory Verbal Hallucinations," *Frontiers in Human Neuroscience* 6 (January 2012): 1.

121 **Broca's area and the auditory cortex:** Judith Ford, "Phenomenology of Auditory Verbal Hallucinations and Their Neural Basis," Hearing Voices: The 2013 Music and Brain Symposium, Stanford University, April 13, 2013, http://www.ustream.tv/recorded/31412393.

122 **"There is Tran, nicknamed Moxi":** Lauren Slater, *Welcome to My Country* (New York: Anchor Books, 1997), 5.

CHAPTER 5: I AM AS IF A DREAM

127 **"How far do our feelings":** Virginia Woolf, *The Letters of Virginia Woolf, Volume 2: 1912–1922.* Nigel Nicolson and Joanne Trautman, eds. (Boston: Houghton Mifflin Harcourt, 1978), 400.

127 **"Forever I shall be":** Albert Camus, *The Myth of Sisyphus and Other Essays* (New York: Vintage, 1991), 19.

131 **"Even though I am":** Quoted in Mauricio Sierra, *Depersonalization: A New Look at a Neglected Syndrome* (Cambridge: Cambridge University Press, 2009), 8.

132 **"An abyss, they say":** Quoted in Ibid.

132 **"a state in which the":** Quoted in Dawn Baker et al., *Overcoming Depersonalization & Feelings of Unreality* (London: Constable and Robinson, 2012), 24.

132 **"I find myself regarding":** Henri-Frédéric Amiel, *Amiel's Journal,* trans. Mary Ward. The Project Gutenberg ebook is at http://www.gutenberg.org/files/8545/8545-h/8545-h.htm.

132 **"whether perception, bodily sensation":** Quoted in Sierra, *Depersonalization,* 17.

133 **"As the car was spinning:"** Russell Noyes Jr. and Roy Kletti, "Depersonalization in Response to Life-Threatening Danger," *Comprehensive Psychiatry* 18, no. 4 (July/August 1977): 375–84.

133 **"The interpretation of depersonalization":** Ibid.

134 **Sarah is a slim:** Some identifying details, including her name, have been changed.

137 **"Of the ideas advanced":** Antonio Damasio, *Self Comes to Mind: Constructing the Conscious Brain* (New York: Vintage Books, 2012), 21.

138 **a term defined by American:** "What is Homeostasis?," *Scientific American,* January 3, 2000, http://www.scientificamerican.com/article/what-is-homeostasis.

138 **"which foreshadows the self":** Damasio, *Self Comes to Mind,* 22.

138 **"broken only by brain":** Ibid.

138 **"provide a direct experience":** Ibid.

139 **"reflect the current state":** Ibid.

140 **"the condition manifests as":** Mauricio Sierra and Anthony S. David, "Depersonalization: A Selective Impairment of Self-Awareness," *Consciousness and Cognition* 20, no. 1 (2011): 99–108.

140 **"(1) feelings of disembodiment":** Lucas Sedeño et al., "How Do You Feel when You Can't Feel Your Body? Interoception, Functional Connectivity and Emotional Processing in Depersonalization-Derealization Disorder," *PLoS One* 9, no. 6 (June 2014): e98769.

141 **suggestive of different neural mechanisms:** Jason J. Braithwaite et al., "Fractionating the Unitary Notion of Dissociation: Disembodied but Not Embodied Dissociative Experiences Are Associated with Exocentric Perspective-Taking," *Frontiers in Human Neuroscience* 7 (October 2013): 1–12.

146 **"contribute to the sense":** Nick Medford, "Emotion and the Unreal Self:

Depersonalization Disorder and De-Affectualization," *Emotion Review* 4, no. 2 (April 2012): 139–44.

147 **"every conceivable kind of feeling":** Damasio, *Self Comes to Mind*, 126.

148 **showed increased activity:** Nick Medford et al., "Functional MRI Studies of Aberrant Self-Experience: Depersonalization Disorder Before and After Treatment," Association for the Scientific Study of Consciousness, http://www.theassc.org/assc15_talks_posters.

149 **"Common sense says":** William James, "What Is an Emotion?" *Mind* 9, no. 34 (1884): 188–205.

149 **"We feel sorry because":** Ibid.

150 **people did not necessarily:** For a complete analysis, see James D. Laird, *Feelings: The Perception of Self* (New York: Oxford University Press, 2007), 65.

151 **"Cognitive factors appear to":** Stanley Schachter and Jerome Singer, "Cognitive, Social and Physiological Determinants of Emotional State," *Psychological Review* 69, no. 5 (September 1962): 379–99.

151 **Beta-blockers essentially inhibit:** Laird, *Feelings*, 72.

152 **"Removing cues from visceral":** Ibid., 73.

152 **the experiments did not take:** Ibid., 78.

154 **The nineteenth-century German physiologist:** Anil Seth, ed., *30-Second Brain* (London: Icon Books, 2014), 50.

154 **The theorem links the conditional:** Anil Ananthaswamy, "I, Algorithm," *New Scientist*, January 29, 2011, 28–31.

158 **when the brain's internal models:** Anil Seth, "Interoceptive Inference, Emotion, and the Embodied Self," *Trends in Cognitive Sciences* 17, no. 11 (November 2013): 565–73.

CHAPTER 6: THE SELF'S BABY STEPS

163 **"Autists are the ultimate square pegs":** Paul Collins, *Not Even Wrong: A Father's Journey into the Lost History of Autism* (New York, London: Bloomsbury, 2004), 225.

163 **"I myself am opaque":** Anne Nesbet, *The Cabinet of Earths* (New York: HarperCollins, 2012), 49.

166 **In 1916, Swiss psychiatrist Paul Eugen:** Uta Frith, ed., *Autism and Asperger Syndrome* (Cambridge: Cambridge University Press, 1991), 6.

166 **"the narrowing of relationships":** Uta Frith, *Autism: Explaining the Enigma*, 2nd ed. (Oxford: Blackwell Publishing, 2003), 5.

167 **"The . . . fundamental disorder is the children's":** Leo Kanner, "Autistic Disturbances of Affective Contact," *Nervous Child* 2 (1943): 217–50.

167 **"There is from the start":** Ibid.

167 **It wasn't until 1980:** "A Cultural History of Autism," PBS, July 29, 2013, http://www.pbs.org/pov/neurotypical/autism-history-timeline.php.

168 **"We must, then, assume ":** Kanner, "Autistic Disturbances." Italics mine.

168 **"The 'Me' corresponds to":** Philippe Rochat, "Emerging Self-Concept," in *The Wiley-Blackwell Handbook of Infant Development*, 2nd ed., J. Gavin Bremner and Theodore D. Wachs, eds. (Oxford: Wiley-Blackwell, 2010), 322.

169 **In 1991, Ulric Neisser:** Ibid., p 323.

169 **Rochat has shown that:** Philippe Rochat and Susan J. Hespos, "Differential Rooting Response by Neonates: Evidence for an Early Sense of Self," *Early Development and Parenting* 6, no. 3–4 (September-December 1997): 105–12.

173 **In 1983, two Austrian psychologists:** Heinz Wimmer and Josef Perner, "Beliefs about Beliefs: Representation and Constraining Function of Wrong Beliefs in Young Children's Understanding of Deception," *Cognition* 13, no. 1 (January 1983): 103–28.

173 **"A travelling salesman":** Ibid.

174 **"beginnings of a capacity":** Alan M. Leslie, "Pretense and Representation: The Origins of 'Theory of Mind,'" *Psychological Review* 94, no. 4 (1987): 412–26.

175 **"It is an early symptom":** Ibid.

175 **"Sally has a basket":** Frith, *Autism: Explaining the Enigma*, 82.

176 **autism involved a specific deficit:** S. Baron-Cohen et al., "Does the Autistic Child Have a 'Theory of Mind'?," *Cognition* 21, no. 1 (October 1985): 37–46.

176 **the three-year-olds forgot:** Alison Gopnik and Janet Astington, "Children's Understanding of Representational Change and Its Relation to the Understanding of False Belief and the Appearance-Reality Distinction," *Child Development* 59, no. 1 (February 1988): 26–37.

178 **"This suggests that these children":** Simon Baron-Cohen, "Are Autistic Children 'Behaviorists'?: An Examination of Their Mental-Physical and Appearance-Reality Distinctions," *Journal of Autism and Developmental Disorders* 19, no. 4 (1989): 579–600. Italics mine.

179 **"freeze the contents":** R. T. Hurlburt et al., "Sampling the Form of Inner Experience in Three Adults with Asperger Syndrome," *Psychological Medicine* 24 (May 1994): 385–95.

179 **"There were no reportable":** Ibid.

181 **Studies have demonstrated strong correlations:** Elizabeth Pellicano, "Links between Theory of Mind and Executive Function in Young Children with Autism: Clues to Developmental Primacy," *Developmental Psychology* 43, no. 4 (July 2007): 974–90.

182 **the right temporoparietal junction:** Hyowon Gweon et al., "Theory of Mind Performance in Children Correlates with Functional Specialization of a Brain Region for Thinking about Thoughts," *Child Development* 83, no. 6 (November/December 2012): 1853–868.

182 **the rTPJ is functionally specialized:** Michael Lombardo et al., "Specialization of Right Temporo-Parietal Junction for Mentalizing and Its Relation to Social Impairments in Autism," *NeuroImage* 56, no. 3 (June 2011): 1832–838.

182 **"the ventromedial prefrontal cortex":** Michael Lombardo et al., "Atypical Neural Self-Representation in Autism," *Brain* 133, no. 2 (February 2010): 611–24.

184 **investigating the French practice:** Laura Spinney, "Therapy for Autistic Children Causes Outcry in France," *The Lancet* 370 (August 2007): 645–46.

184 **"alleged therapy":** David Amaral et al., "Against *Le Packing*: A Consensus Statement," *Journal of the American Academy of Child & Adolescent Psychiatry* 50, no. 2 (February 2011): 191–2.

185 **John closer to his body:** Angèle Consoli et al., "Lorazepam, Fluoxetine and Packing Therapy in an Adolescent with Pervasive Developmental Disorder and Catatonia," *Journal of Physiology—Paris* 104, no. 6 (September 2010): 309–14.

185 **"combine the body and":** David Cohen et al., "Investigating the Use of Packing Therapy in Adolescents with Catatonia: A Retrospective Study," *Clinical Neuropsychiatry* 6, no. 1 (2009): 29–34.

185 **"to reinforce children's consciousness":** Ibid.

186 **"based on clinical observations":** Elizabeth B. Torres et al., "Autism: The Micro-Movement Perspective," *Frontiers in Integrative Neuroscience* 7 (July 2013): 1–26.

187 **"There could be error feedback":** Ian P. Howard and Brian J. Rogers, *Perceiving in Depth, Volume 3: Other Mechanisms of Depth Perception* (Oxford: Oxford University Press, 2012), 266.

190 **"there is a high probability":** Karl Friston, "The Free-Energy Principle: A Unified Brain Theory?" *Nature Reviews Neuroscience* 11 (February 2010): 127–38.

190 **"resist a natural tendency":** Ibid.

190 **"Biological agents must avoid":** Ibid.

191 **"A magical world suggests":** Pawan Sinha et al., "Autism as a Disorder of Prediction," *Proceedings of the National Academy of Sciences* 111, no. 42 (October 2014): 15220–5225.

192 **"Theory of mind is inherently":** Ibid.

CHAPTER 7: WHEN YOU ARE BESIDE YOURSELF

195 **"This proposition [that]":** René Descartes, "Meditations on First Philosophy," trans. Elizabeth S. Haldane, in *The Philosophical Works of Descartes* (Cambridge: Cambridge University Press, 1911), 9.

195 **"'Owning' your body":** Thomas Metzinger, *The Ego Tunnel: The Science of the Mind and the Myth of the Self* (New York: Basic Books, 2009), 75.

197 **he jumped out:** Peter Brugger et al., "Heautoscopy, Epilepsy, and Suicide," *Journal of Neurology, Neurosurgery & Psychiatry* 57, no. 7 (1994): 838–39.

198 **"No . . . no . . . of course not":** Guy de Maupassant, *The Horla*, trans. Charlotte Mandell (New York: Melville House, 2005), 41.

199 **Eliot was inspired:** Sunil Kumar Sarker, *T. S. Eliot: Poetry, Plays and Prose* (New Delhi: Atlantic, 2000), 103.

199 **"I know that during":** Sir Ernest Shackleton, *South! The Story of Shackleton's Last Expedition (1914–1917)*, available at http://www.gutenberg.org/files/5199/5199-h/5199-h.htm.

202 **mentioned in an article:** Constance Holden, ed., "Doppelgängers," *Science* 291 (January 19, 2001): 429.

202 **"perhaps the most famous":** Nicholas Wade, "Guest Editorial," *Perception* 29 (2000): 253–57.

202 **"parts of my body":** G. M. Stratton, "Some Preliminary Experiments on Vision without Inversion of the Retinal Image," *Psychological Review* 3 (1896): 611–17.

202 **published another paper:** G. M. Stratton, "The Spatial Harmony of Touch and Sight," *Mind* 8 (October 1899): 492–505.

203 **"In the more languidly":** Ibid.

204 **The strength of the illusion:** H. Henrik Ehrsson et al., "That's My Hand! Activity in Premotor Cortex Reflects Feeling of Ownership of a Limb," *Science* 305 (August 6, 2004): 875–77.

205 **temperature in the real hand:** G. Lorimer Moseley et al., "Psychologically Induced Cooling of a Specific Body Part Caused by the Illusory Ownership of an Artificial Counterpart," *Proceedings of the National Academy of Sciences* 105, no. 35 (September 2008): 13169–3173.

205 **Ehrsson's team:** Arvid Guterstam et al., "The Invisible Hand Illusion: Multisensory Integration Leads to the Embodiment of a Discrete Volume of Empty Space," *Journal of Cognitive Neuroscience* 25, no. 7 (July 2013): 1078–1099.

209 **Metzinger's OBEs stopped:** For more details, see Metzinger, *The Ego Tunnel.*

210 **leading to the woman's OBE:** Olaf Blanke et al., "Stimulating Illusory Own-Body Perceptions," *Nature* 419 (September 19, 2002): 269–70.

211 **In the synchronous condition:** Bigna Lenggenhager et al., "Video Ergo Sum: Manipulating Bodily Self-Consciousness," *Science* 317, no. 5841 (August 24, 2007): 1096-1099.

211 **"looking at my own body":** Silvio Ionta et al., "Multisensory Mechanisms in Temporo-Parietal Cortex Support Self-Location and First-Person Perspective," *Neuron* 70, no. 2 (April 2011): 363–74.

212 **yet another full-body illusion:** Valeria I. Petkova and H. Henrik Ehrsson, "If I Were You: Perceptual Illusion of Body Swapping," *PLoS One* 3, no. 12 (December 2008): e3832.

213 **correlation being strongest:** Valeria I. Petkova et al., "From Part- to Whole-Body Ownership in the Multisensory Brain," *Current Biology* 21 (July 12, 2011): 1118–122.

218 **Heydrich and Blanke:** Lukas Heydrich and Olaf Blanke, "Distinct Illusory Own-Body Perceptions Caused by Damage to Posterior Insula and Extrastriate Cortex," *Brain* 136 (2013): 790–803.

219 **the *minimal phenomenal self*:** Olaf Blanke and Thomas Metzinger, "Full-Body Illusions and Minimal Phenomenal Selfhood," *Trends in Cognitive Sciences* 13, no. 1 (2009): 7–13.

222 **the phenomenon of *mineness*:** Jakob Hohwy, "The Sense of Self in the Phenomenology of Agency and Perception," *Psyche* 13, no. 1 (April 2007): 1–20.

223 **"One's own body size":** Björn van der Hoort et al., "Being Barbie: The Size of One's Own Body Determines the Perceived Size of the World," *PLoS One* 6, no. 5 (May 2011): e20195.

224 **were less able to recall:** Loretxu Bergouignan et al., "Out-of-Body–Induced Hippocampal Amnesia," *Proceedings of the National Academy of Sciences* 111, no. 12 (March 2014): 4421–426.

CHAPTER 8: BEING NO ONE, HERE AND NOW

225 **"If the doors of perception":** William Blake, *The Marriage of Heaven and Hell,* available at http://www.gutenberg.org/files/45315/45315-h/45315-h.htm.

225 **"I feel a happiness":** Quoted in Jacques Catteau, *Dostoyevsky and the Process of Literary Creation* (Cambridge: Cambridge University Press, 1989), 114.

229 **"as if I had lost":** Shirley M. Ferguson Rayport, "Dostoyevsky's Epilepsy: A New Approach to Retrospective Diagnosis," *Epilepsy & Behavior* 22, no. 3 (2011): 557–70.

229 **"The sensation is so strong":** Quoted in Catteau, *Dostoyevsky,* 114.

229 **"a moment or two":** Fyodor Dostoyevsky, *The Idiot,* trans. Eva Martin, available at http://www.gutenberg.org/files/2638/2638-h/2638-h.htm.

229 **"I feel then as":** Ibid.

229 **"if when I recall":** Ibid.

230 **"I believe that the":** Henri Gastaut, "Fyodor Mikhailovitch Dostoevski's Involuntary Contribution to the Symptomatology and Prognosis of Epilepsy," *Epilepsia* 19, no. 2 (1978): 186–201.

230 **"He says that the pleasure":** F. Cirignotta et al., "Temporal Lobe Epilepsy with Ecstatic Seizures (So-called Dostoevsky Epilepsy)," *Epilepsia* 21 (1980): 705–10.

231 **"During the seizure it":** Fabienne Picard and A. D. Craig, "Ecstatic Epileptic Seizures: A Potential Window on the Neural Basis for Human Self-Awareness," *Epilepsy & Behavior* 16, no. 3 (2009): 539–46.

232 **"a sensation of velvet":** Ibid.

232 **"The immense joy that":** Ibid.

239 **She got news of:** Anil Ananthaswamy, "Fits of Rapture," *New Scientist,* January 25, 2014, 44.

239 **"I don't believe the":** A. D. Craig, "How Do You Feel?," http://vimeo.com/8170544.

241 **"The thermal grill reveals":** A. D. Craig, "Can the Basis for Central Neuropathic Pain Be Identified by Using a Thermal Grill?," *Pain* 135, no. 3 (April 2008): 215–16.

241 **the experience of pain:** A. D. Craig et al., "Functional Imaging of an Illusion of Pain," *Nature* 384 (November 21, 1996): 258–60.

241 **In his subsequent studies:** A. D. Craig et al., "Thermosensory Activation of Insular Cortex," *Nature Neuroscience* 3, no. 2 (February 2000): 184–90.

242 **"Activation of the ACC":** A. D. Craig, "How Do You Feel? Interoception: The Sense of the Physiological Condition of the Body," *Nature Reviews Neuroscience* 3 (August 2002): 655–66.

242 **"It seems to provide":** A. D. Craig, "Interoception and Emotion: A Neuroanatomical Perspective," in *Handbook of Emotions,* 3rd ed., Michael Lewis et al., eds. (New York: Guilford Press, 2008), 281.

242 **"the material self as":** Ibid., 281.

242 **"source of the sense":** Antonio Damasio, "Mental Self: The Person Within," *Nature* 423 (May 15, 2003): 227.

244 **But when the electrode:** Fabienne Picard et al., "Induction of a Sense of Bliss by Electrical Stimulation of the Anterior Insula," *Cortex* 49, no. 10 (2013): 2935–937.

245 **"One bright May morning":** Aldous Huxley, *The Doors of Perception* (London: Thinking Ink, 2011), 2.

245 **"relish the possibility":** "Dr Humphry Osmond," *Telegraph*, February 16, 2004, http://www.telegraph.co.uk/news/obituaries/1454436/Dr-Humphry-Osmond.html.

245 **"At breakfast that morning":** Huxley, *The Doors of Perception,* 5.

246 **"Space was still there":** Ibid., 7.

246 **"To fathom hell or soar angelic":** Obituary of Humphry Osmond in *BMJ* 328 (March 20, 2004): 713.

246 **Neuroimaging studies of people:** Franz X. Vollenweider and Michael Kometer, "The Neurobiology of Psychedelic Drugs: Implications for the Treatment of Mood Disorders," *Nature Reviews Neuroscience* 11 (September 2010): 642–51.

246 **In one double-blind study:** Jordi Riba et al., "Increased Frontal and Paralimbic Activation Following Ayahuasca, the Pan-Amazonian Inebriant," *Psychopharmacology* 186, no. 1 (2006): 93–98.

247 **the anterior insula might:** A. D. Craig, "How Do You Feel—Now? The Anterior Insula and Human Awareness," *Nature Reviews Neuroscience* 10 (January 2009): 59–70.

248 **In 2006, Martin Paulus and Murray Stein:** Martin P. Paulus and Murray B. Stein, "An Insular View of Anxiety," *Biological Psychiatry* 60, no. 4 (August 2006): 383–87.

248 **Picard posits that the opposite:** Fabienne Picard, "State of Belief, Subjective Certainty and Bliss as a Product of Cortical Dysfunction," *Cortex* 49, no. 9 (October 2013): 2494–500.

249 **"joy, creativity, the process":** Mihaly Csikszentmihalyi, *Flow: The Psychology of Optimal Experience* (New York: Harper Perennial, 2008), xi.

249 **"One item that disappears":** Ibid. 62.

249 **"the optimal experience":** Ibid., 64.

249 **loss of self-consciousness:** Ibid.

EPILOGUE

254 **Both synchronic and diachronic:** Matthew R. Dasti, "Nyāya," *Internet Encyclopedia of Philosophy,* http://www.iep.utm.edu/nyaya.

256 **"I am not the mind":** Personal communication, translation from Sanskrit by C. S. Aravinda, TIFR Centre for Applicable Mathematics, Bangalore, India.

257 **"When I enter most intimately":** David Hume, *A Treatise of Human Nature,* available at http://www.gutenberg.org/files/4705/4705-h/4705-h.htm.

257 **"Each normal individual":** Daniel C. Dennett, *Consciousness Explained* (Boston: Little Brown, 1991), 416.

257 **"is the same kind of thing":** Daniel C. Dennett, *Intuition Pumps and Other Tools for Thinking* (New York: W. W. Norton, 2013), 334.

257 **"fiction, posited in order":** Ibid., 336.

257 **"entirely fabricated from":** Miri Albahari in Mark Siderits et al., eds., *Self, No Self? Perspectives from Analytical, Phenomenological, & Indian Traditions* (Oxford: Oxford University Press, 2010), 92.

258 **"When the brain manages":** Antonio Damasio, *Self Comes to Mind: Constructing the Conscious Brain* (New York: Vintage, 2012), 11.

258 **"The self is introduced":** John R. Searle, "The Mystery of Consciousness Continues," review of Damasio's *Self Comes to Mind*, *New York Review of Books*, June 9, 2011, http://www.nybooks.com/articles/archives/2011/jun/09/mystery-consciousness-continues.

261 **"mistake lies in taking":** Siderits et al., eds., *Self, No Self?*, 23.

261 **"our mistake lies":** Ibid., 23.

INDEX

INDEX

ABOUT THE AUTHOR

Anil Ananthaswamy is former deputy news editor and current consultant for *New Scientist*. He is a guest editor at UC Santa Cruz's renowned science-writing program and teaches an annual science journalism workshop at the National Centre for Biological Sciences in Bangalore, India. He is a freelance feature editor for the Proceedings of the National Academy of Sciences' "Front Matter" and has written for *National Geographic News*, *Discover*, and *Matter*. He has been a columnist for PBS NOVA's *The Nature of Reality* blog. He won the UK Institute of Physics' Physics Journalism Prize and the Association of British Science Writers' award for Best Investigative Journalism. His first book, *The Edge of Physics*, was voted book of the year in 2010 by *Physics World*. He lives in Bangalore, India, and Berkeley, California.